The Global Tech Revolution
to End Human Trafficking

Deebo Haven Center Presents

The Global Tech Revolution to End Human Trafficking

*How AI, Blockchain, Drones, and Emerging Technologies
Are Dismantling Modern Slavery*

By Damien Brinson

Founder & Executive Director, Deebo Haven Center
Orlando, Florida

© 2026 Deebo Haven Center

All rights reserved.

For permission requests, contact:

Email: WeCare@deebohavencenter.org

Website: https://www.deebohavencenter.org

ISBN (Print): 979-8-9953931-0-8

Printed in the United States of America

First Edition: 2026

A Note on the Stories in This Book

The personal stories shared throughout this book are fictional composites created to illustrate the realities of human trafficking. They are inspired by documented patterns, published research, and the collective experiences shared by survivors and advocates worldwide. No character represents a specific individual. All names are invented. These stories are included to honor the truth of what millions endure, while protecting the privacy and dignity of real survivors.

Foreword

From darkness into light: a survivor's threshold

This foreword is a composite voice, inspired by the experiences of trafficking survivors who have gone on to work in technology and advocacy. It does not represent any single individual.

I was seventeen when I disappeared.

Not in the way the movies show it, with screeching tires and a van door sliding shut. I disappeared slowly, one lie at a time, one locked door at a time, one stolen choice at a time. By the time I understood what was happening to me, I had already been sold twice, moved three times, and convinced by my trafficker that no one was looking for me.

He was wrong about that. Someone was looking. But the systems that were supposed to find me, the databases, the hotlines, the cross-agency reports, could not keep up with the speed at which I was being moved. I was in a different city every ten days. My trafficker changed my name, my phone, my appearance. The paper trail was designed to end before it began.

I was eventually recovered, not by a traditional investigation, but because an AI system flagged an online advertisement that matched a pattern associated with trafficking. The algorithm detected linguistic markers in the post, cross-referenced the phone number against a network of known trafficking ads, and generated an alert that reached the right detective at the right time. Within forty-eight hours, I was out.

A machine saw me when the world could not.

I share this not to celebrate technology as a savior. Technology did not hold my hand in the hospital. Technology did not sit with me through the nightmares or teach me how to trust again. People did that. Counselors. Advocates. Survivors who had walked the same road before me and came back to light it for others. Technology is a tool. What matters is who wields it, and why.

But I also share this because I know what it feels like to be invisible in a system that was not designed to find you. And I know the difference it makes when someone builds a system that is.

After my recovery, I struggled. The world expected me to be grateful to be alive, and I was, but gratitude does not pay rent. It does not erase a criminal record that was created under a name that was not mine. It does not undo the years of education I lost or the credit history I never had.

What changed my trajectory was digital literacy. A nonprofit program taught me how to use a computer, really use one. Not just social media, but encrypted communication, online safety, resume building, basic coding. For the first time in my life, I felt like I had power that no one could take from me. The same internet that had been used to exploit me became the tool I used to rebuild.

Today, I work as a technology consultant for an anti-trafficking organization. I advise engineers on how to design tools that respect survivor autonomy. I review AI systems for bias. I train law enforcement on what the digital experience of trafficking actually looks like from the inside. And I mentor other survivors who are finding their way into careers in technology.

When Damien Brinson asked me to write this foreword, I hesitated. Not because I did not believe in the book. I had read the manuscript, and it is the most comprehensive, honest, and accessible work I have seen on how technology is being deployed against trafficking. What made me hesitate was the weight of representing a community that is so often spoken about but so rarely spoken to.

Survivors are not statistics. We are not case studies. We are not the "before" picture in someone else's success story. We are people who survived something that was designed to destroy us, and many of us have spent the years since then building expertise, resilience, and vision that the anti-trafficking movement desperately needs.

This book honors that. It does not sensationalize. It does not exploit. It tells the truth about how trafficking works, how technology is fighting it, and most importantly, how every person reading these pages can contribute to the solution. It treats the reader as an intelligent partner in the fight, not a passive observer.

If you are a technologist, this book will show you what your skills can do beyond profit margins.

If you are a policymaker, it will arm you with the knowledge to write smarter, bolder legislation.

If you are a survivor, it will remind you that the world is changing, and your voice is part of the reason why.

If you are a student, a parent, a teacher, or simply someone who believes that the world can be better, this book will show you the tools that are already making it so, and it will ask you to pick one up.

I used to be invisible. Technology helped someone see me. Now I spend my life making sure the next generation of tools sees everyone.

Read this book. Then do something with what you learn.

The world is waiting for you to show up.

A Survivor, An Advocate, A Builder

For the ones who were told they were invisible.

For the ones who survived what no one should ever have to endure.

For the ones still waiting to be found.

This book is your proof that someone is fighting.

You were never invisible. You were always worth finding.

Table of Contents

Foreword

Introduction

Part I: Technology Against Trafficking

Part II: Inside the Machine

Part III: Final Reflections and Action

Introduction

A luminous globe of interconnected hope

Human trafficking is one of the darkest stains on humanity: an invisible war that claims millions of lives across every continent, every community, and every demographic. It doesn't discriminate. It festers in hidden corners of the internet, thrives in unstable economies, and flourishes under the radar in both developing nations and first-world cities that pride themselves on progress.

I need you to hear that again.

This is not happening somewhere else. It is happening here. In your city. In your supply chain. On your children's screens.

For the past several years, I have been on the front lines of this fight, serving as the Founder and Executive Director of the Deebo Haven Center, a nonprofit based in Orlando, Florida, committed to the rescue, rehabilitation, and reintegration of survivors of human trafficking. Through this work, I've encountered heartbreaking stories, labyrinthine systems of exploitation, and a global trafficking industry that is more technologically advanced than most people realize.

That's why I wrote this book.

While the media often focuses on dramatic rescues or criminal prosecutions, the conversation rarely includes the transformative power of technology, especially emerging technologies like AI, blockchain, drones, and virtual reality, to disrupt trafficking at scale. We hear about the problem. Rarely do we hear about the arsenal being built to destroy it.

This book is a response to that gap. It is designed not only for nonprofit leaders and law enforcement professionals, but also for technologists, policymakers, students, and everyday advocates who want to understand how innovation can be wielded for liberation. It is both a call to action and a roadmap forward.

Why This Book Was Written

As someone deeply embedded in both anti-trafficking advocacy and organizational leadership, I've seen firsthand the limitations of conventional tools. Survivors need faster intervention. Law enforcement needs smarter intelligence. Advocates need stronger infrastructure. And society as a whole needs a clearer vision of what's possible when technology meets human compassion.

But here's the tension: the same technologies that can save lives can also be weaponized to exploit them. Traffickers are not behind the curve, they are early adopters. They use encrypted platforms to sell human beings. They use cryptocurrency to launder blood money. They use social media algorithms to groom children. And they use the dark web as an invisible marketplace where people are bought and sold like inventory.

If traffickers are adapting with technology, so must we. Not reluctantly. Not eventually. Now.

This book brings together the most promising innovations in one cohesive volume. It is grounded in research, supported by real-world case studies, and driven by a belief that the same digital revolution that empowered exploitation can be repurposed, redirected, to end it.

What This Book Covers

Each chapter explores a unique frontier where technology is intersecting with the global movement to end trafficking. Here's your roadmap:

Chapter 1: The Dark Web and Trafficking. Unpacks how traffickers exploit encrypted platforms and how AI and surveillance tools are now being used to infiltrate underground networks and intercept illegal activity.

Chapter 2: Artificial Intelligence in Pattern Recognition.
Explores how AI detects trafficking indicators from online ads,
social media posts, financial transactions, and movement
patterns, empowering faster victim identification and response.

Chapter 3: Blockchain for Transparency and Accountability.
Demonstrates how blockchain is being used to track supply
chains, secure victim records, trace illicit transactions, and
enforce ethical accountability.

**Chapter 4: Big Data and Its Power to Disrupt Human
Trafficking.** Highlights the use of large-scale data analytics to
map trafficking trends, predict hotspots, and support multi-
agency intelligence sharing in real time.

Chapter 5: AI-Powered Surveillance and Image Recognition.
Shows how AI-driven surveillance and facial recognition
systems are helping law enforcement detect trafficking in
airports, hotels, border crossings, and beyond.

**Chapter 6: Virtual and Augmented Reality in Education and
Prevention.** Explains how immersive tech is being used to train
officers, educate the public, simulate real-life trafficking
experiences, and support survivor recovery.

**Chapter 7: Drones and Satellite Technology in Rescue
Operations.** Details how unmanned aerial vehicles and satellite
imaging are locating hidden victims, monitoring remote
trafficking zones, and guiding rescue teams with precision.

**Chapter 8: Collaboration Between Tech Companies and
NGOs.** Outlines the growing alliances between nonprofits and
tech firms, and how these partnerships are producing
groundbreaking tools, databases, and reporting systems.

Chapter 9: Legal and Ethical Considerations in Using AI to Combat Trafficking. Tackles the essential concerns of data privacy, algorithmic bias, survivor consent, and international legal standards, ensuring that innovation does not come at the cost of justice.

Chapter 10: The Future of Technology in the Fight Against Human Trafficking. Casts a forward-looking vision for how AI, quantum computing, wearable tech, and predictive modeling could eradicate trafficking altogether, if deployed ethically and collaboratively.

My Hope for This Book

More than anything, I hope this book serves as both a guide and an invitation.

A guide for those looking to build smarter systems, launch stronger coalitions, and create scalable, lasting change.

An invitation to those with technological talent, capital, or influence to step into the arena, to use their skills not just for profit, but for protection. Not just for innovation, but for liberation.

Survivors deserve more than reactive justice. They deserve proactive prevention. And with the right tools, deployed by the right people, with the right values, we can give them exactly that.

The fight against human trafficking is not just a moral obligation. It's a technological challenge. And it's one that we are now, finally, equipped to win.

This book is my contribution to that fight.

For the victims still waiting to be found.
For the survivors still waiting to be heard.
And for the world we are all called to help build,
where freedom is non-negotiable, and dignity is digital by design.

Let's begin.

Damien Brinson

Founder & Executive Director, Deebo Haven Center

Orlando, Florida

Part I

Technology Against Trafficking

Chapter 1: The Dark Web and Trafficking

Light piercing the encrypted tunnel

Amara was fourteen when she first heard the promise that would change everything.

It came through a direct message on a social media platform she used to share dance videos with friends in Lagos. The sender's profile picture showed a smiling woman in a business suit. The message was simple: a modeling opportunity in Europe. Good pay. A visa. A new life. Amara's mother had been sick for months, and the family was drowning in medical debt. The offer felt like oxygen.

Within two weeks, Amara was on a plane. Within a month, she was locked in an apartment in a city she couldn't name, her passport confiscated, her phone destroyed. The "modeling agency" did not exist, at least not on the surface web. The transaction that sold her had been negotiated on an encrypted forum buried deep in the dark web, paid for in cryptocurrency that left no trace in any bank.

Amara's story is not unique. It is replicated thousands of times a day, across continents, across languages, across every demographic you can imagine. And increasingly, the invisible infrastructure that makes it possible lives in the same place: the dark web.

The Hidden Layers of the Internet

Most people interact with what technologists call the surface web: the indexed, searchable internet where you read the news, check your email, and scroll social media. It accounts for roughly four to five percent of the internet's total content. Beneath that lies the deep web: password-protected databases, academic archives, medical records, and corporate intranets that search engines cannot reach.

And beneath that, hidden behind layers of encryption and accessible only through specialized software like Tor (The Onion Router), lies the dark web.

Not everything on the dark web is illicit. In censored or repressive environments, reporters, dissidents, and whistleblowers rely on anonymous networks to share information and avoid surveillance. Yet the same privacy protections that shield legitimate users also attract criminal markets. For trafficking networks, that makes the dark web an efficient place to recruit, coordinate, and profit.

To understand how technology can dismantle trafficking, we first have to understand where it hides. The dark web is ground zero.

How Traffickers Weaponize the Dark Web

Traffickers have adopted the dark web not out of curiosity but out of strategic necessity. It offers them three things that are existential to their operations: anonymity, encrypted communication, and untraceable financial transactions. Let me walk you through how each one works in practice.

Recruitment through deception. The journey for many victims begins on the surface web, through social media, dating apps, or job boards, but the coordination happens deeper. Traffickers use encrypted messaging services hosted on the dark web to communicate with recruiters, share dossiers on targets, and coordinate logistics. A fraudulent job posting in Nairobi connects to a handler in Istanbul connects to a buyer on a dark web forum in Western Europe. The victim never sees the chain. The investigators struggle to trace it.

Marketplaces for human beings. It is difficult to write these words, but they must be written: dark web marketplaces exist that function like e-commerce platforms for trafficking. Victims' physical descriptions, ages, and photographs are listed. Buyers browse. Prices are set or auctioned. The platforms operate with user ratings, escrow systems, and customer service: a grotesque mirror of legitimate online commerce. These sites cycle rapidly, shutting down and re-emerging under new names to evade law enforcement.

Cryptocurrency laundering. Traffickers use Bitcoin, Monero, and other cryptocurrencies to conduct transactions. While Bitcoin's blockchain records every transaction publicly, the identities behind wallets remain pseudonymous. Monero goes further, offering privacy features that make transactions nearly impossible to trace. Mixing services and chain-hopping, moving funds between multiple cryptocurrency networks, add additional layers of obfuscation. The result is a financial pipeline that moves blood money across borders without touching a single bank.

Exploitation content as currency. The dark web is also a distribution hub for child sexual abuse material (CSAM) and live-streamed exploitation. These materials serve multiple purposes for traffickers: they generate revenue through pay-per-view models, they are used as leverage to blackmail victims into compliance, and they attract networks of buyers who may escalate from consuming content to purchasing victims directly. The emergence of AI-generated CSAM, synthetic images created by generative AI tools trained on real abuse material, has added a devastating new dimension, flooding platforms with content that is harder to trace to specific victims while still fueling demand and normalizing exploitation.

Fighting Back: Technology Enters the Dark Web

If the dark web is the fortress, then the counter-trafficking community is building the siege engines. And they are getting more sophisticated every year.

AI-powered dark web crawlers. Organizations like Thorn, a nonprofit co-founded by Ashton Kutcher and Demi Moore, have developed AI tools that crawl dark web forums and marketplaces, scanning for language patterns, image hashes, and behavioral indicators associated with trafficking. Thorn's Spotlight platform uses machine learning to analyze online ads and identify victims of sex trafficking, processing data at a speed and scale no human team could match. Since its launch,

Spotlight has assisted law enforcement in identifying thousands of victims, many of them minors.

Natural language processing for coded communication. Traffickers speak in code. They use slang, emoji sequences, and euphemisms that evolve constantly to evade detection. Natural language processing (NLP) algorithms are now trained to detect these linguistic fingerprints, analyzing not just keywords but context, tone, and patterns of communication across forums and messaging platforms. When a new code phrase emerges in one market, NLP systems can flag it across the entire dark web within hours.

Blockchain forensics. While cryptocurrencies offer pseudonymity, they are not invisible. Companies like Chainalysis and Elliptic have developed blockchain analytics platforms that trace the flow of cryptocurrency across wallets, exchanges, and mixing services. By mapping transaction patterns and linking wallet addresses to known entities, these tools have helped law enforcement agencies seize millions of dollars in illicit assets and dismantle major trafficking operations. In recent years, joint operations between Europol and the FBI have increasingly relied on blockchain forensics to trace trafficking-linked cryptocurrency payments across multiple countries, leading to the dismantling of networks and the seizure of millions of dollars in illicit assets.

International cooperation platforms. The dark web does not respect borders, and neither can the response. Interpol's Cyber Fusion Centre and Europol's European Cybercrime Centre (EC3) facilitate real-time intelligence sharing between law enforcement agencies worldwide. Through platforms like the International Child Sexual Exploitation (ICSE) database, investigators from over sixty-eight countries can cross-reference images, share leads, and coordinate operations at a speed that was unimaginable a decade ago.

AI-generated decoys and sting operations. One of the most promising, and ethically complex, developments is the use of AI-generated decoy profiles to bait traffickers. Law enforcement agencies have begun deploying chatbots that mimic potential victims on dark web forums, engaging traffickers in conversations that reveal their networks, methods, and locations. These digital decoys can operate around the clock without risk to human operatives, gathering evidence that has led to successful prosecutions.

The Generative AI Threat

We cannot talk about the dark web in 2026 without confronting the elephant in the room: generative AI.

The same artificial intelligence that can write code, compose music, and generate photorealistic images is now being weaponized by traffickers and predators. Deepfake technology allows the creation of synthetic nude images or videos of real people, including minors, without their knowledge or consent. AI-generated CSAM has exploded across dark web platforms, creating a volume of abusive content that threatens to overwhelm detection systems designed to match known images.

Traffickers are using AI chatbots to automate grooming at scale, engaging dozens of potential victims simultaneously with personalized manipulation that adapts in real time. AI voice cloning allows scammers to impersonate family members or authority figures, adding a terrifying layer of deception to recruitment schemes.

The counter-trafficking community is responding. Organizations like the National Center for Missing & Exploited Children (NCMEC), the Internet Watch Foundation (IWF), and the Tech Coalition are developing AI classifiers that can detect synthetic content, even when it doesn't match any known image hash. Hash-sharing databases like PhotoDNA are being upgraded to identify AI-generated material. And policymakers are beginning to close legal loopholes: the DEFIANCE Act and the TAKE IT DOWN Act in the United States, along with the EU's AI Act, are among the first legislative efforts to criminalize AI-generated exploitation content and mandate platform accountability.

But the race is on. And right now, the offense has the advantage.

The New Frontier: Gaming Platforms and the Metaverse

There is a recruitment channel that is almost entirely absent from anti-trafficking literature, and it is one of the fastest-growing digital spaces on the planet: online gaming.

Hundreds of millions of minors log into multiplayer gaming platforms every day. Fortnite, Roblox, Minecraft, Discord servers, and dozens of other platforms offer built-in chat, voice communication, and private messaging. For children and teenagers, these platforms feel safe. They are playing games with friends. They are building worlds. They are having fun.

Traffickers know this. And they are there.

Predators use gaming platforms to build relationships with minors over weeks or months, establishing trust through shared gameplay before moving the conversation to private channels. The grooming follows a predictable arc: flattery, gift-giving (in-game currency or real money), isolation from friends and family, introduction of sexual content, and eventually coercion or recruitment. By the time the child realizes what is happening, the emotional manipulation is deeply entrenched.

The metaverse, including virtual reality social platforms like VRChat and Horizon Worlds, adds a new dimension of risk. In immersive virtual environments, users interact through avatars that can touch, gesture, and simulate physical proximity. Reports of harassment and grooming in VR spaces have surged, and the line between virtual and real exploitation becomes dangerously thin when a child's avatar is being manipulated by an adult who is building toward real-world contact.

What exists today. Some platforms have implemented AI-powered chat moderation that flags suspicious language patterns in real time. Xbox and PlayStation have parental control systems that limit communication with strangers. Thorn has developed tools that detect grooming language in chat environments. Discord has partnered with NCMEC to report exploitation content through the CyberTipline.

What is missing. There is no industry-wide standard for child safety in gaming. Age verification remains trivially easy to circumvent. Voice chat, which is the primary communication channel in many games, is far more difficult to monitor than text. And VR environments present entirely new challenges for detecting exploitation, because the interactions are embodied and spatial rather than textual.

The policy vacuum. Current child protection laws were not written for virtual worlds. Questions that have no clear legal answers include: Is avatar-on-avatar sexual harassment a crime? Can grooming that occurs entirely within a game be prosecuted? Who is liable when a platform designed for children is used for exploitation? These questions demand urgent legislative attention, and the gaming industry must be at the table.

Here is what I need you to understand as we close this chapter and prepare to go deeper:

The dark web is not a distant underworld.
It is the backroom of the internet you use every day.
And inside it, right now, someone is being sold.

But also inside it, right now, an algorithm is scanning. A crawler is mapping. A blockchain trail is being followed. A decoy is gathering evidence. And someone, maybe someone reading this book, is building the next tool that will make the dark web a little less dark.

Amara was eventually identified through facial recognition technology that matched an image from a dark web listing to a missing persons database. She was recovered in a joint operation between Interpol and local law enforcement, guided by intelligence that started with a single flagged cryptocurrency transaction.

She was one of the lucky ones. Millions are still waiting.

The question this book keeps asking, the question every chapter will bring you back to, is this:

What will you build?
What will you fund?
What will you demand?

The technology exists. The darkness is not stronger than the light we are building. But the light only wins if we aim it.

Questions for Reflection and Discussion

1. How does the existence of trafficking marketplaces on the dark web change your understanding of online safety, both for yourself and for vulnerable populations?

2. Generative AI creates both new threats (deepfakes, synthetic CSAM) and new tools for detection. How should society balance the open development of AI with the risk of its misuse?

3. What role should gaming platforms and metaverse companies play in protecting minors from grooming and recruitment? Where does corporate responsibility end and parental responsibility begin?

4. If you could fund one technology to combat dark web trafficking, what would it be and why?

Let's keep going.

Chapter 2: Artificial Intelligence in Pattern Recognition

A constellation of detected patterns

The ad appeared on a classified website in the Midwest, sandwiched between furniture listings and used cars. It advertised "young companionship" with a local phone number and a photograph of a girl who looked barely seventeen. To the average browser, it might have been invisible, just another post in an ocean of digital noise.

But to Spotlight, the AI platform developed by Thorn, the ad was a flare.

The algorithm had flagged it within minutes of posting. Not because of a single keyword, but because of a constellation of signals: the phrasing matched linguistic patterns documented across thousands of known trafficking advertisements. The phone number was linked to six other ads in three different states, all posted within the same seventy-two-hour window. The image metadata revealed it had been taken in a hotel room whose carpet pattern matched a major chain. And the girl in the photo, her name was later confirmed as Maya, had been reported missing from a group home in Ohio nine days earlier.

Maya was recovered within forty-eight hours of the ad being flagged. Her trafficker was arrested with evidence compiled almost entirely by artificial intelligence.

This is what pattern recognition looks like when it works. This is AI in the service of liberation.

The Science of Seeing What Humans Miss

Human trafficking leaves traces. Not always obvious ones, traffickers are sophisticated, adaptive, and deliberate in covering their tracks, but traces nonetheless. A cluster of online ads with recycled phone numbers. A series of financial transactions that follow the same unusual cadence. A social media account that suddenly shifts from personal posts to coded language. A face that appears in surveillance footage at three different airports in a single week.

The problem has never been the absence of clues. It has been the sheer volume of data in which those clues are buried. A single analyst reviewing online classified ads could spend weeks finding what an AI system identifies in seconds. A financial investigator tracking suspicious transactions across global banking systems might take months to uncover what machine learning models detect in real time.

Artificial intelligence doesn't replace human investigators. It gives them superpowers. It processes the noise and surfaces the signal. It turns haystacks into needles.

Machine Learning: Teaching Computers to Think Like Investigators

At the core of AI pattern recognition is machine learning, algorithms that improve through experience. These systems are trained on historical data: thousands of confirmed trafficking cases, verified advertisements, known financial patterns, and documented behavioral indicators. The more data they consume, the sharper they become.

Predictive recruitment detection. Machine learning models trained on trafficking cases can identify high-risk online job advertisements, social media interactions, and forum posts that match known recruitment patterns. An unusual frequency of "high-paying international work" postings targeting young women in economically vulnerable regions, for example, triggers automatic escalation. These systems don't just find individual ads, they map networks, linking recruiters across platforms and geographies to reveal operations that span continents.

Geospatial hotspot prediction. AI tools powered by geographic information systems (GIS) analyze historical trafficking data alongside economic indicators, migration patterns, conflict zones, and even seasonal trends to predict where trafficking is likely to spike next. Before a major international sporting event, these models can identify the neighborhoods most at risk and help law enforcement deploy resources preemptively rather than reactively. This is the difference between rescuing someone and preventing them from needing rescue in the first place.

Financial anomaly detection. Banks and financial institutions process billions of transactions daily. AI systems monitor these flows in real time, flagging patterns that deviate from established baselines: rapid cash movements between shell accounts, unusual cryptocurrency conversions, or payments that match known laundering typologies. Major financial institutions, including HSBC, JPMorgan Chase, and Western Union, now deploy AI-driven anti-trafficking transaction monitoring, and the Financial Action Task Force (FATF) has recommended AI integration as a global best practice.

Natural Language Processing: Decoding the Language of Exploitation

Traffickers are linguists of evasion. They invent slang, rotate code words, and use emoji sequences that change weekly. A skull emoji might mean a threatened victim in one network and available inventory in another. The word "fresh" in a classified ad has an entirely different meaning in a trafficking context than in a grocery store.

Natural language processing (NLP) is the branch of AI that teaches machines to understand human language, not just words, but meaning, intent, and deception. NLP systems trained on trafficking communication can detect subtle linguistic markers that human reviewers would miss, especially at scale.

Text and keyword analysis across platforms. NLP algorithms scan millions of posts across chatrooms, dating apps, encrypted messaging platforms, and classified ad sites. They analyze not just individual keywords but clusters of language, contextual phrasing, and communication cadence. When a new code word emerges in one trafficking market, NLP systems can detect its spread across the ecosystem within days.

Sentiment and coercion detection. Beyond keywords, NLP can analyze tone. Grooming conversations follow predictable emotional arcs, building trust, creating dependency, introducing threats. NLP models trained on these patterns can flag conversations that are progressing toward exploitation, potentially intervening before a victim is fully ensnared.

In one landmark initiative, NLP analysis of online escort advertisements identified a set of hidden linguistic signals that traffickers used to indicate the availability of minors. The patterns were so subtle, variations in punctuation, specific emoji combinations, deliberate misspellings, that human analysts had missed them for years. Once the AI identified the code, law enforcement agencies across multiple jurisdictions were able to coordinate rescues that recovered dozens of underage victims.

Real-Time Detection: AI That Never Sleeps

One of AI's greatest strengths in anti-trafficking work is its ability to operate continuously, monitoring data streams twenty-four hours a day, seven days a week, without fatigue, bias, or emotional burnout.

Social media and image analysis. Traffickers use social media platforms, Instagram, TikTok, Snapchat, Facebook, to recruit and control victims. AI monitoring systems scan publicly accessible posts, stories, and images for indicators of exploitation: coercive language in comments, images showing signs of captivity or duress, and accounts that suddenly shift from personal content to commercial sexual advertisement. Facial recognition technology cross-references images against databases of missing and exploited persons, generating alerts when matches are found.

Real-time financial surveillance. AI systems deployed by major banks operate as continuous sentinels over global financial networks. When a transaction pattern matches a known trafficking typology, say, a series of small deposits from multiple prepaid cards into a single account, followed by an immediate wire transfer to an overseas shell company, the system flags it, pauses it, and escalates it to human investigators within minutes.

Consider the case of a multinational banking AI that detected an anomalous pattern: identical dollar amounts being wired from fourteen different cities to a single account in Eastern Europe every Thursday at the same time. Human investigators, guided by the AI's analysis, discovered a trafficking network that was moving victims between cities on a weekly rotation, with each wire transfer representing payment for a different victim's "services." The network was dismantled. The victims were recovered. The AI had seen what no single analyst could have.

The Arms Race: Challenges and the Road Ahead

For every advance in AI-powered detection, traffickers adapt. They change their language. They switch platforms. They fragment their operations into smaller, harder-to-detect cells. This is an arms race, and pretending otherwise would be dishonest.

Data privacy and civil liberties. Using AI to monitor communications and financial transactions raises legitimate concerns about surveillance overreach. The same tools that detect trafficking can, if unregulated, be used to monitor political dissidents, journalists, or ordinary citizens. Strong legal frameworks, independent oversight, and transparency requirements are essential guardrails.

Algorithmic bias. AI systems are only as fair as the data they're trained on. If training datasets overrepresent certain demographics or geographies, the resulting models may produce biased outcomes, disproportionately flagging vulnerable communities while missing trafficking in affluent settings. Continuous auditing, diverse training data, and human oversight are non-negotiable.

Adversarial adaptation. Traffickers study the tools used against them. When they learn that an AI system flags certain keywords, they change the keywords. When they discover that blockchain forensics can trace Bitcoin, they switch to Monero or privacy coins. The counter-trafficking community must invest not just in current tools but in adaptive AI, systems that learn and evolve as fast as the threat does.

The future is promising. Explainable AI (XAI) will allow investigators to understand, and courts to trust, the reasoning behind algorithmic alerts. Quantum computing will accelerate data processing by orders of magnitude. And global data-sharing frameworks will enable AI systems to learn from trafficking patterns across every continent simultaneously, creating a collective intelligence that no single network can outrun.

Artificial intelligence is not a silver bullet. It is a force multiplier.

It takes the dedication, the expertise, and the moral courage of human investigators and amplifies them to a scale that matches the problem. Maya was not saved by an algorithm alone, she was saved by people who built the algorithm, people who deployed it, and people who knocked on a hotel room door at two in the morning to bring her home.

AI finds the pattern.

Humans break the chain.

Together, they are unstoppable.

In the next chapter, we move from detection to accountability, exploring how blockchain technology is creating transparent, immutable records that traffickers cannot erase, alter, or hide behind.

The digital breadcrumbs are everywhere. The question is whether we're building the systems to follow them.

Questions for Reflection and Discussion

1. AI pattern recognition can flag potential trafficking activity, but false positives can harm innocent people. How should investigators balance sensitivity with accuracy?

2. Should financial institutions be legally required to implement AI-based trafficking detection, or should it remain voluntary?

3. How can the anti-trafficking community ensure that AI systems do not disproportionately target marginalized communities while missing trafficking in affluent settings?

4. What safeguards should be in place before AI-generated alerts are used to justify law enforcement action?

Chapter 3: Blockchain for Transparency and Accountability

The chain that lifts, the ledger that remembers

Kofi was twelve when he was taken from his village in Ghana and put to work on a fishing boat on Lake Volta. For three years, he dove into murky water to untangle nets, worked eighteen-hour days, and ate whatever scraps the boat owner decided to give him. He had no documents. No proof of identity. No birth certificate. When a rescue team finally reached him, they faced a problem almost as cruel as the trafficking itself: without official identification, Kofi did not officially exist.

He could not enroll in school. He could not access healthcare. He could not be placed with family, because there was no verified record connecting him to anyone. In the bureaucratic systems designed to protect children, Kofi was invisible: a ghost in the machine of a world that had already failed him once.

Now imagine a different scenario. Imagine that Kofi's identity had been registered at birth on a blockchain-based digital ID system, a decentralized, tamper-proof record that no government collapse, no corrupt official, and no trafficker could erase. Imagine that when the rescue team scanned a simple QR code, Kofi's name, birthdate, family connections, and medical history appeared instantly, verified, secure, and unalterable.

That technology exists today. It is being piloted in refugee camps, deployed in supply chains, and tested in recovery programs. And it has the potential to transform not just how we rescue survivors, but how we prevent trafficking from happening in the first place.

This is the promise of blockchain.

Blockchain Explained: The Ledger That Cannot Lie

Blockchain technology was originally developed as the infrastructure behind Bitcoin, but its applications have expanded far beyond cryptocurrency. At its core, a blockchain is a decentralized digital ledger: a chain of records (called blocks) that are linked together cryptographically. Each block contains a set of transactions or data entries, a timestamp, and a reference to the previous block. Once a block is added to the chain, it cannot be altered or deleted without invalidating every subsequent block.

Three properties make blockchain uniquely powerful for anti-trafficking work:

Decentralization. No single entity controls the ledger. Data is distributed across a network of computers, making it nearly impossible for any individual, including a corrupt official or a trafficker, to manipulate records.

Immutability. Once data is recorded, it is permanent. A victim's identity, a supply chain audit, a financial transaction, once on the blockchain, they cannot be erased, backdated, or falsified.

Transparency with privacy. Blockchain can be designed to be publicly verifiable while keeping sensitive details encrypted. Authorized parties can access the information they need without exposing a victim's personal data to the world.

These properties create a foundation of trust in systems where trust has historically been absent, exactly the kind of environments where trafficking thrives.

Digital Identity for the Invisible

Recent World Bank identity data indicate that roughly 800 million people still lack official proof of identity. Without documentation, access to school enrollment, banking, healthcare, legal remedies, and formal employment can disappear. That absence of recognized identity does not just create inconvenience; it creates vulnerability that traffickers routinely exploit.

Stateless populations, refugees, undocumented migrants, and children born outside formal systems are disproportionately vulnerable to trafficking precisely because they have no official record of existence. Blockchain-based digital identity systems are beginning to change that.

Self-sovereign identity. Platforms like the ID2020 Alliance, Hyperledger Indy, and Microsoft's ION network are developing decentralized identity solutions that give individuals control over their own digital identity. A refugee fleeing conflict can carry a verified digital ID on a mobile device: an ID that no border guard can confiscate and no smuggler can destroy.

Continuity of care for survivors. When a trafficking survivor is recovered, their journey through rehabilitation involves multiple agencies: law enforcement, medical providers, legal services, shelters, and reintegration programs. Blockchain can create a continuous, secure record that travels with the survivor, ensuring they receive coordinated care without having to retell their traumatic story at every new intake. The record is encrypted, access-controlled, and verifiable, protecting the survivor's dignity while ensuring accountability among service providers.

Supply Chain Transparency: Following the Thread

Human trafficking is not only about individual victims. It is embedded in the global economy. The shirt you're wearing, the phone in your hand, the chocolate in your pantry, each may carry the fingerprints of forced labor. The International Labour Organization estimates that forced labor generates $236 billion in illegal profits annually, with exploitation concentrated in agriculture, manufacturing, mining, construction, and domestic work.

The complexity of modern supply chains, with raw materials passing through dozens of intermediaries across multiple countries before reaching a consumer, makes it extraordinarily difficult to verify that no exploitation occurred along the way. Blockchain is changing the calculus.

Provenance tracking. Companies like Provenance, Everledger, and IBM Food Trust are using blockchain to create end-to-end records of product journeys. Every handoff, from raw material to manufacturer to distributor to retailer, is recorded on an immutable ledger. If a garment factory in Bangladesh claims its workers are fairly paid, the blockchain record either confirms or contradicts that claim with auditable data.

Smart contracts for ethical compliance. Smart contracts are self-executing agreements coded directly onto a blockchain. A company could embed anti-trafficking compliance requirements into its supplier contracts: if a third-party audit detects labor violations, the smart contract automatically suspends payment and triggers an investigation. No human intervention needed. No loopholes. No "we didn't know."

Major corporations are beginning to adopt these systems. Nestlé has invested in traceability initiatives for its cocoa supply chain, including pilot programs that use digital tracking to map the journey of raw materials from farm to factory. De Beers uses the Tracr platform to trace diamonds from mine to retail, ensuring they are conflict-free. And the Responsible Business Alliance, a coalition of over two hundred electronics companies, is exploring blockchain-based labor compliance monitoring across global manufacturing networks.

Following the Money: Blockchain Forensics

Traffickers love cryptocurrency because they believe it makes them invisible. They're wrong, and blockchain is the reason why.

Every Bitcoin transaction is recorded permanently on a public ledger. While the wallets are pseudonymous, the transactions themselves are transparent. Blockchain analytics firms like Chainalysis and Elliptic have developed sophisticated tools that trace the flow of funds across wallets, exchanges, and mixing services, building evidence trails that have led to some of the most significant trafficking prosecutions in recent history.

In 2021, a joint operation between the FBI, Europol, and Australian Federal Police used blockchain analytics to dismantle a major dark web child exploitation network. The investigation, part of a broader international effort that included operations like the takedown of the DarkMarket marketplace, followed cryptocurrency payments through over a dozen wallets, identified key operators, and led to arrests across multiple countries. Hundreds of children were safeguarded as a result.

Privacy coins like Monero present a greater challenge, offering built-in obfuscation that makes tracing more difficult. But the forensics community is adapting. Academic researchers and private-sector analysts are developing probabilistic tracing methods that can identify Monero transactions with increasing confidence. The arms race continues, but the trajectory is clear: the idea that cryptocurrency provides perfect anonymity is a myth that blockchain forensics is systematically dismantling.

A Shared Ledger for a Shared Fight

One of the most persistent obstacles in anti-trafficking work is fragmentation. Law enforcement in one country may have intelligence that would break a case in another, but incompatible systems, jurisdictional barriers, and trust deficits prevent information from flowing where it's needed. Blockchain offers a solution.

A decentralized, permissioned blockchain could serve as a global trafficking intelligence ledger: a shared database accessible to vetted law enforcement agencies, NGOs, and international organizations. Each entry would be cryptographically verified, timestamped, and immutable. Intelligence from a rescue in Thailand could instantly inform an investigation in Germany. A pattern detected in one jurisdiction could trigger alerts across the network.

Organizations like the Global Modern Slavery Directory and STOP THE TRAFFIK are already building blockchain-informed data-sharing platforms. The technology is ready. What's needed is the political will to deploy it at scale.

Kofi's story did not end at the lake. After his rescue, an NGO working with blockchain-based identity systems created a digital record that connected him to his birth village, verified his age, and enabled his enrollment in school. Three years later, he was one of the top students in his district.

A blockchain entry didn't save Kofi. People did. But the blockchain ensured that once he was saved, the system could never lose him again.

Transparency is the enemy of exploitation.
Immutability is the antidote to erasure.
And blockchain is the ledger that refuses to forget.

In the next chapter, we zoom out even further, exploring how Big Data is mapping the entire trafficking landscape, predicting where exploitation will strike next, and enabling responses that are not just faster but smarter.

Questions for Reflection and Discussion

1. How could blockchain-based digital identity systems change the lives of the estimated 800 million people worldwide who lack official identification?

2. As a consumer, would you pay more for a product that uses blockchain-verified ethical supply chains? What would make that information trustworthy to you?

3. What are the privacy risks of storing survivor data on a blockchain, and how can those risks be mitigated while maintaining the benefits of immutability?

4. How could smart contracts be designed to automatically enforce anti-trafficking labor standards in global supply chains?

Chapter 4: Big Data and Its Power to Disrupt Human Trafficking

Data streams converging on the signal

The first time Priya tried to tell someone what was happening to her, she typed three words into a search engine on a borrowed phone: "help I'm trapped."

She was nineteen, recruited from a village in Rajasthan with promises of domestic work in Mumbai. Instead, she found herself locked in a garment factory with forty other women, sewing fourteen hours a day, sleeping on the floor, earning nothing. Her documents had been taken. Her family had been told she was dead.

That search query, three desperate words, was captured anonymously in a dataset analyzed by a research team studying digital distress signals in labor trafficking hotspots. It became one data point among thousands. Alone, it meant nothing. Combined with location data, factory registration records, shipping manifests, and labor complaint databases, it became part of a pattern that led investigators to the factory and the liberation of forty-three women.

This is the power of Big Data. Not any single piece of information, but the convergence of millions of data points into a picture so detailed that trafficking networks can no longer hide in plain sight.

The Data Deluge: Volume, Velocity, and Variety

Human trafficking investigations now sit inside a constant torrent of digital information. Social posts, payment records, satellite imagery, mobile metadata, search behavior, travel logs, and camera feeds generate a volume of evidence no human team can review manually. The challenge is no longer whether clues exist; it is whether institutions can assemble and interpret them fast enough.

Big Data is defined by three characteristics that make it both overwhelming and extraordinarily powerful: volume (the sheer amount), velocity (the speed at which it's generated), and variety (the range of sources and formats). When applied to human trafficking, these characteristics transform raw information into actionable intelligence.

Traditional investigative methods, tip lines, informants, manual surveillance, remain important. But they are inherently limited by human bandwidth. A detective can review a hundred classified ads in a shift. A Big Data system can analyze a hundred million in the same time, cross-referencing each against financial records, geographic data, and known trafficking indicators.

Mapping the Invisible: Online Activity and Digital Footprints

Traffickers operate in digital spaces because that's where their victims are. Job boards, social media platforms, dating apps, encrypted messaging services, these are the hunting grounds. Big Data tools scrape, aggregate, and analyze these digital environments at scale.

Recruitment pattern detection. By analyzing millions of online job advertisements, Big Data algorithms can identify postings that match known trafficking recruitment profiles: vague job descriptions, unusually high pay for low-skill work, emphasis on travel and relocation, and targeting of specific vulnerable demographics. These flags don't confirm trafficking, but they prioritize where human investigators should look first.

Behavioral shift analysis. Big Data can track changes in online behavior that may indicate someone is being groomed or controlled. A social media account that suddenly goes quiet after a period of frantic messaging. A phone number that begins appearing in multiple escort advertisements across different cities. A user who searches for help resources in a language different from their usual online activity. Each of these signals, isolated, might mean nothing. Together, they paint a portrait that investigators can act on.

Partnerships between technology companies and organizations like the Counter-Trafficking Data Collaborative (CTDC), maintained by the International Organization for Migration, have used Big Data analytics to monitor classified ad platforms across Southeast Asia and other high-risk regions. These systems identify networks of accounts using nearly identical language and posting schedules across multiple countries, revealing trafficking operations that recruit women under the guise of hospitality and domestic work. The pattern-matching capabilities of Big Data have contributed to the identification and recovery of victims who would otherwise have disappeared into exploitation.

Following the Money at Scale

If data is the new oil, financial data is the refinery where trafficking's profits are processed. Big Data analytics applied to global financial systems have become one of the most effective tools for identifying and disrupting trafficking operations.

Transaction pattern analysis. AI-powered Big Data systems monitor billions of financial transactions daily, flagging anomalies that match known trafficking typologies. Sudden, large cash deposits followed by immediate international wire transfers. Clusters of prepaid card purchases at the same location. Cryptocurrency movements that follow the timing patterns of known exploitation networks.

The Financial Crimes Enforcement Network (FinCEN) in the United States has published advisory guidance encouraging financial institutions to use Big Data analytics for trafficking detection. Banks like HSBC, JPMorgan Chase, and Standard Chartered have integrated AI-driven monitoring systems that specifically target trafficking-related financial patterns, resulting in thousands of Suspicious Activity Reports (SARs) filed annually.

Priya's factory was eventually linked to a financial pattern: small, regular payments from a network of international buyers, routed through a series of shell companies to the factory owner's personal accounts. Big Data connected the dots across three countries and two currencies in a matter of days, work that would have taken human investigators months.

Predictive Analytics: Seeing Tomorrow's Trafficking Today

Perhaps the most transformative application of Big Data is its ability to predict trafficking before it happens.

Predictive analytics uses historical data, past trafficking incidents, economic indicators, migration patterns, conflict data, weather events, and even social media sentiment, to forecast where and when trafficking is likely to occur. These models don't replace human judgment. They augment it, giving law enforcement and prevention organizations a head start.

Event-driven prediction. Major international events, sporting championships, political summits, large festivals, have historically been accompanied by spikes in trafficking activity. Predictive models can analyze data from previous events to identify the specific neighborhoods, venues, and online platforms most likely to be targeted, enabling preemptive deployment of resources.

Climate and conflict modeling. Natural disasters and armed conflicts displace populations, creating pools of vulnerable people that traffickers exploit. Big Data models that integrate climate projections, conflict monitoring, and displacement tracking can identify at-risk populations before traffickers reach them. After Typhoon Haiyan devastated the Philippines in 2013, trafficking surged in affected areas. Today, similar models would trigger early-warning alerts and preventive interventions.

The Global Estimates of Modern Slavery, a landmark report series produced jointly by the International Labour Organization, Walk Free, and the International Organization for Migration, aggregates trafficking and forced labor data from over 160 countries, creating a comprehensive picture of global exploitation patterns. Alongside platforms like the Counter-Trafficking Data Collaborative, these resources enable researchers and policymakers to allocate resources, design interventions, and track progress: a Big Data infrastructure for the entire anti-trafficking movement.

The Challenge of Data

Big Data is powerful, but it is not infallible. And the ethical challenges are as significant as the analytical ones.

Privacy versus protection. Analyzing the digital footprints of millions of people in search of trafficking patterns inevitably captures the data of innocent individuals. Anonymization protocols, data minimization principles, and strict access controls are essential, not optional, but foundational.

Data quality and bias. Big Data systems are only as reliable as their inputs. Incomplete datasets, underreporting in certain regions, overrepresentation of certain trafficking types, can produce skewed results. A model trained primarily on sex trafficking data may be blind to labor trafficking indicators, missing Priya's factory entirely.

Interoperability. The anti-trafficking landscape is fragmented. NGOs, law enforcement agencies, financial institutions, and tech companies each hold pieces of the puzzle, often in incompatible formats and siloed systems. Building the infrastructure for secure, standardized data sharing is as important as building the algorithms that analyze it.

Priya is free now. She works with an anti-trafficking organization in Mumbai, helping other survivors navigate the recovery process. She often says the same thing when she tells her story:

> *"I was one search query in a billion. But someone built a system that could hear me."*

That system was Big Data. And it is getting louder.

The data exists.
The patterns are there.
The question is whether we build the systems to listen.

In the next chapter, we shift from data analysis to physical surveillance, exploring how AI-powered cameras, facial recognition, and image analysis are transforming the ability to spot trafficking in the real world, in real time.

Questions for Reflection and Discussion

1. Predictive analytics can forecast trafficking hotspots, but predictions are not certainties. How should law enforcement act on probabilistic intelligence without overreacting?

2. What ethical frameworks should govern the collection and analysis of personal data for anti-trafficking purposes?

3. How can Big Data initiatives ensure they include data from underreported regions and trafficking types, rather than reinforcing existing biases?

4. If a major international event is predicted to increase trafficking activity, what specific preventive measures should organizers, governments, and NGOs implement?

Chapter 5: AI-Powered Surveillance and Image Recognition

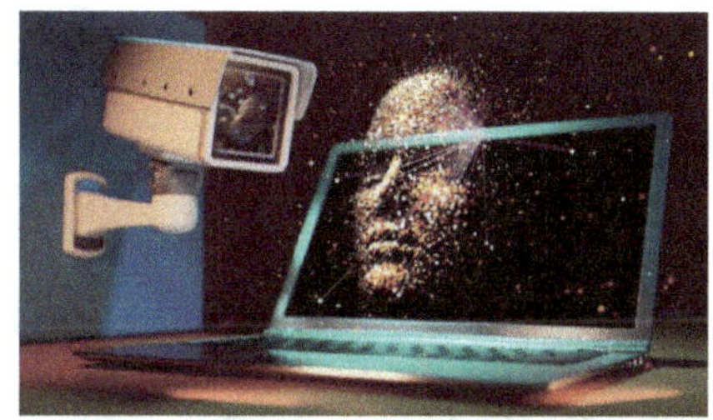

The watchful lens of protection

The security camera at Gate B7 of a major European airport recorded thousands of faces every hour. Most were business travelers, families on vacation, students heading to study abroad. The footage scrolled endlessly across a monitor that a security officer glanced at between sips of coffee.

But one face, at 3:47 a.m. on a Tuesday, triggered something no human eye would have caught.

Her name was Lena. She was sixteen. She had been reported missing in Romania three weeks earlier. The AI system running behind the airport's surveillance network matched her face to a missing persons database in under two seconds. It flagged the man walking beside her, his hand clamped around her wrist, her eyes cast downward, and cross-referenced him against a law enforcement watchlist. He had two prior trafficking convictions under different names.

Border police intercepted them at the boarding gate. Lena was on a flight to a country where she had no contacts, no resources, and no way home. In forty-five minutes, she would have been gone.

AI saw her. AI stopped it.

The Evolution of Surveillance

Surveillance has long been part of policing, but legacy systems are reactive by design. They rely on operators to notice the right face, gesture, or pattern at exactly the right moment across a wall of screens. In trafficking cases, where subtle cues matter and footage volumes are enormous, that dependence on sustained human attention creates blind spots traffickers can exploit.

AI-powered surveillance fundamentally changes the equation. Computer vision algorithms, deep learning neural networks, and image recognition systems can analyze thousands of video feeds simultaneously, identifying objects, behaviors, and faces with a speed and accuracy that human operators cannot match. These systems don't get tired. They don't look away. And they are increasingly being deployed in the specific locations where trafficking victims are most likely to be found.

Seeing Trafficking in Real Time

Behavioral analysis in public spaces. AI video analytics can be trained to recognize behavioral patterns associated with trafficking: a person who appears to be under the control of another, individuals moving through restricted areas with signs of distress, groups arriving at hotels with characteristics that match known exploitation patterns. The system doesn't make arrests, it generates alerts that direct human attention to where it matters most.

Hotels have become a critical front line. The TraffickCam app, developed by the Exchange Initiative, allows everyday travelers to upload photos of their hotel rooms. AI systems compare these images against the backgrounds of photos recovered from trafficking advertisements, matching carpet patterns, bedspreads, and wall fixtures to specific hotel locations. This crowdsourced approach has provided law enforcement with location intelligence that has contributed to numerous investigations.

Facial recognition at scale. Modern facial recognition systems can process millions of faces against databases of missing persons, known traffickers, and persons of interest. At transportation hubs, airports, bus stations, train terminals, border crossings, these systems create a digital net that traffickers must pass through. When deployed with appropriate legal safeguards, facial recognition has proven to be one of the most effective tools for identifying victims in transit.

India's TrackChild system, integrated with the national Crime and Criminal Tracking Network (CCTNS), uses facial recognition to match unidentified children with missing persons reports. Since its deployment, the system has helped trace thousands of missing children, many of whom were trafficking victims.

Image hash matching and CSAM detection. Beyond facial recognition, AI systems use perceptual hashing, mathematical fingerprints of images, to identify known child sexual abuse material across platforms. Technologies like Microsoft's PhotoDNA and Meta's PDQ create unique hashes for each known CSAM image, enabling automatic detection even when images are cropped, filtered, or resized. The Tech Coalition's Project Protect is building shared infrastructure to extend these capabilities across the industry.

Movement Tracking and Geographic Intelligence

AI surveillance doesn't just identify individual faces, it maps movement patterns over time and space.

Digital trail mapping. By tracking individuals across multiple camera networks, AI can construct movement profiles that reveal trafficking routes. A person photographed at a bus station in one city, then at a hotel in another city three hundred miles away the next day, then at a third location the following week, creates a movement pattern that AI can flag as consistent with trafficking logistics.

Crowd analysis. AI systems can analyze crowd dynamics in public spaces, identifying anomalies such as groups of young people who appear to be escorted by a single older individual, individuals who seem disoriented or unable to communicate freely, or patterns of movement that suggest coerced transit.

Lena's trafficker had been recorded at the same airport six times in two months, each time with a different young woman. The AI system had built his movement profile over weeks before the alert that saved Lena. When investigators pulled the full picture, they discovered a pipeline that had moved an estimated dozen victims across European borders.

The Ethics of Watching

I will not pretend that AI surveillance is without risk. It is not. And anyone who tells you otherwise is selling you a world without complexity.

Privacy and civil liberties. Deploying facial recognition and behavioral analysis in public spaces raises profound questions about the balance between safety and freedom. Mass surveillance has been used by authoritarian regimes to suppress dissent, monitor minorities, and control populations. The same technology that identifies trafficking victims can, in the wrong hands, identify political activists.

Algorithmic bias in facial recognition. Multiple studies, including landmark research by Joy Buolamwini at the MIT Media Lab, have demonstrated that facial recognition systems are significantly less accurate for women and people of color. False positive rates are higher for these demographics, creating the risk of wrongful detention, missed identifications, and discriminatory enforcement.

Accountability frameworks. AI surveillance in anti-trafficking work must be governed by clear legal frameworks that include: judicial oversight of deployment, strict data retention limits, mandatory accuracy audits, transparent reporting on outcomes, and independent review mechanisms. The European Union's AI Act, which entered into force in August 2024 with provisions phasing in through 2026, classifies real-time biometric surveillance as "high-risk" and imposes stringent requirements on its use: a model that other jurisdictions are beginning to follow.

The goal is not surveillance for its own sake. The goal is targeted, accountable, rights-respecting deployment of technology in service of the most vulnerable.

The Hotel and Hospitality Technology Gap

Hotels are among the most common venues for sex trafficking. Victims are moved through hotel rooms on a nightly or weekly rotation, and the signs are often visible to anyone paying attention: short-duration stays paid in cash, multiple rooms booked under the same name, a steady flow of visitors to a single room, a young person who appears frightened or controlled by a companion. Yet the technology deployed in the hospitality industry to detect these patterns is shockingly underdeveloped.

AI-powered check-in anomaly detection. A handful of hotel chains have begun piloting AI systems that analyze booking patterns for trafficking indicators. The algorithms flag reservations that match known exploitation typologies: rooms booked for a few hours at unusual times, multiple rooms on the same floor reserved simultaneously, cash or prepaid card payments that avoid identity verification. When a pattern is flagged, hotel staff receive a discreet alert to observe and, if warranted, contact law enforcement.

TraffickCam and crowdsourced room identification. The TraffickCam app, developed by the Exchange Initiative, invites travelers to upload photographs of hotel rooms wherever they stay. These images are matched against the backgrounds of photos recovered from online trafficking advertisements. Carpet patterns, bedspreads, wall art, and bathroom fixtures become forensic evidence that can place a trafficking victim in a specific hotel. The database now contains millions of images and has contributed to dozens of investigations.

Smart room sensors. Emerging IoT technology could enable hotel rooms to detect anomalous activity without invading guest privacy. Sensors that monitor door-open frequency, noise levels, or room occupancy patterns could flag situations consistent with exploitation, such as a room with constant visitor turnover or a door opening and closing dozens of times in a single night, without recording audio, video, or identifying information.

Staff training through VR. As discussed in the previous chapter, virtual reality training simulations place hotel staff in realistic trafficking scenarios, teaching them to recognize the signs and respond appropriately. The American Hotel and Lodging Association (AHLA) has partnered with anti-trafficking organizations to develop training programs, but adoption across the industry remains uneven.

The hospitality industry has a unique opportunity, and a unique responsibility, to become a frontline defense against trafficking. The guests walking through lobbies and booking rooms online are, in some cases, trafficking victims. The technology to identify them exists. What is needed is the will to deploy it across the industry, not as a public relations initiative, but as an operational standard.

Lena is back in Romania now, living with her grandmother and attending school. She does not know the name of the algorithm that flagged her face. She does not know about the server farms or the neural networks or the hash-matching databases.

She knows she was seen.

In a world that made her invisible, a machine looked at a screen and said: *she matters.*

That is what ethical surveillance looks like.
Not a camera watching everyone.
But a system designed to find the ones who are lost.

In the next chapter, we step into entirely new territory, virtual and augmented reality, and explore how immersive technology is reshaping how we train, educate, and heal in the fight against trafficking.

Questions for Reflection and Discussion

1. Facial recognition can save lives but also threatens privacy. Where do you draw the line between protective surveillance and invasive monitoring?

2. How should the hospitality industry be held accountable for implementing trafficking detection technology in hotels?

3. Joy Buolamwini's research showed that facial recognition is less accurate for women and people of color. What steps must be taken before these systems are deployed in anti-trafficking contexts?

4. Would you support mandatory AI surveillance in airports and transportation hubs if it could identify trafficking victims? What conditions would you require?

Chapter 6: Virtual and Augmented Reality in Education and Prevention

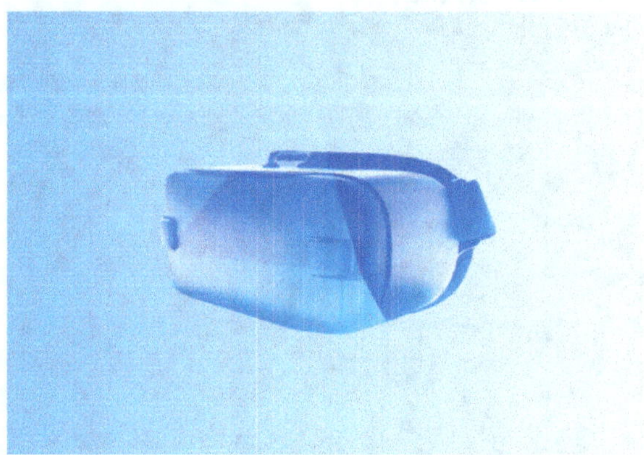

Immersive worlds that teach what textbooks cannot

Officer Daniels had been on the force for twelve years. He'd responded to domestic violence calls, drug raids, and one hostage situation. He considered himself experienced, steady under pressure. Then he put on a VR headset.

The simulation placed him in a motel room. A young woman sat on the edge of a bed, arms crossed, eyes averted. A man, her "boyfriend," he said, stood by the door, answering every question she was asked. She flinched when he moved. She called him "Daddy." When Officer Daniels asked her directly if she needed help, her eyes darted to the man before she whispered, "No."

In the debrief, Daniels was quiet for a long time. Then he said: "I've been on calls like that. I missed it. I missed the signs because I didn't know what I was looking at."

He never missed them again.

Beyond the Textbook: Why Immersive Technology Matters

Human trafficking is a crime that hides in plain sight. Victims are coached to deny their situation. Traffickers present as romantic partners, employers, or family members. The signs of exploitation are subtle: a particular kind of fear, a rehearsed script, a body language that screams for help while the mouth says everything is fine.

Traditional training methods, slide decks, case studies, written protocols, can teach the theory. But they cannot replicate the experience. They cannot put a trainee in the room and let them feel the tension, read the body language, and make the split-second judgment call that determines whether a victim is identified or walks back out the door with her trafficker.

Virtual reality can.

VR immerses users in fully simulated environments where they interact with AI-driven characters, make decisions, and experience consequences in real time. Augmented reality overlays digital information onto the real world, enhancing situational awareness with data, alerts, and contextual intelligence. Together, these technologies are transforming anti-trafficking education, training, prevention, and even survivor recovery.

Training That Changes Behavior

Law enforcement preparedness. Police departments, border agencies, and social workers across Europe, North America, and Australia have begun integrating VR-based training into their anti-trafficking programs. These simulations recreate trafficking scenarios with remarkable fidelity, from the layout of the room to the micro-expressions on a victim's face, and train officers to identify indicators that textbook training alone cannot convey.

Research from the University of Southern California's Institute for Creative Technologies has shown that VR-based training improves threat recognition, empathetic response, and decision-making speed compared to traditional methods. Officers who complete VR trafficking simulations are significantly more likely to correctly identify trafficking victims during real-world encounters.

Healthcare provider recognition. Emergency room doctors, nurses, and clinic staff are often the only professionals who interact with trafficking victims outside the control of their traffickers. VR training programs place healthcare workers in simulated clinical encounters where they must recognize the signs of exploitation, unexplained injuries, patients who defer to companions, reluctance to provide personal information, and respond appropriately.

The SOAR (Stop, Observe, Ask, Respond) training program, developed by the U.S. Department of Health and Human Services, has incorporated VR elements to enhance its effectiveness. Healthcare providers report that the immersive experience fundamentally shifts their awareness in ways that lectures and pamphlets do not.

Empathy at Scale: Public Awareness Through Immersion

One of the greatest barriers to anti-trafficking action is the empathy gap. Most people know trafficking exists in the abstract. Few understand what it feels like, how it operates, or how close it is to their daily lives.

Immersive tools narrow that distance between abstract knowledge and lived reality.

Empathy-building simulations. Organizations including the United Nations, the International Justice Mission, and several universities have developed VR experiences that place participants in the perspective of trafficking victims. Users experience the recruitment process, the moment of entrapment, the daily reality of exploitation, and the complexity of escape, all in a safe, controlled environment.

These experiences are not entertainment. They are designed to provoke reflection, understanding, and action. Studies show that participants who complete VR empathy simulations donate more, volunteer more, and engage more actively with anti-trafficking causes than those who receive the same information through traditional media.

AR awareness campaigns. Augmented reality is being deployed in public spaces to deliver anti-trafficking information where it's most needed. AR-enabled posters in airports, bus stations, and rest areas allow travelers to scan a code with their smartphone and instantly access information about trafficking indicators, emergency hotlines, and reporting resources. Campaigns in India, the UK, and the United States have used AR to dramatically increase public engagement with anti-trafficking messaging.

Healing Through Virtual Worlds

The applications of VR extend beyond prevention and into recovery. Trafficking survivors often carry profound trauma, post-traumatic stress disorder, anxiety, depression, and dissociation. Traditional therapy is essential, but VR offers a complementary tool that is showing remarkable results.

Virtual reality exposure therapy (VRET). Under the guidance of trained therapists, survivors can revisit elements of their traumatic experiences in a controlled virtual environment. This gradual, supported exposure helps reduce the intensity of trauma responses over time, allowing survivors to process their experiences without being overwhelmed. Clinical studies have shown that VRET can significantly reduce PTSD symptoms in trauma survivors.

Life skills rebuilding. Survivors who spent months or years in captivity often lack the practical skills needed for independent living. VR training modules simulate everyday situations, job interviews, grocery shopping, public transportation, financial management, allowing survivors to practice in a low-pressure environment before facing them in reality. These programs restore confidence and accelerate reintegration.

AR in the Field: Real-Time Intelligence for First Responders

Augmented reality is not just an educational tool, it is becoming an operational one.

AR-enhanced field operations. Prototype systems equip law enforcement officers and border agents with AR glasses that display real-time data during field operations: facial recognition alerts, database lookups, suspect profiles, and victim information overlaid directly onto the officer's field of view. This eliminates the delay of radio communication and manual database searches, enabling faster, more informed decision-making at the point of contact.

AR navigation for rescue operations. In complex or unfamiliar environments, warehouses, industrial compounds, dense urban areas, AR can project navigational guidance and structural information directly into rescuers' visual field, improving operational efficiency and officer safety during trafficking raids.

Officer Daniels now leads his department's anti-trafficking unit. He requires every new officer to complete the VR simulation before their first field assignment. "You can't unsee it," he says. "And once you've seen it, really seen it, you don't miss the signs anymore."

Technology doesn't just change what we know.
It changes what we notice.
And noticing is where rescue begins.

In the next chapter, we take to the skies, exploring how drones and satellite technology are transforming rescue operations in the most remote and dangerous places on earth.

Questions for Reflection and Discussion

1. Could VR empathy simulations change public attitudes toward trafficking in ways that traditional media cannot? What makes immersive experiences different?

2. What ethical guidelines should govern the use of VR therapy with trafficking survivors, particularly regarding informed consent and potential retraumatization?

3. How could AR technology be deployed in your community to raise awareness about trafficking?

4. Should VR-based trafficking recognition training be mandatory for professionals in law enforcement, healthcare, hospitality, and education?

Chapter 7: Drones and Satellite Technology in Rescue Operations

Eyes in the sky, justice on the ground

The camp was invisible from the road. Forty kilometers from the nearest town, hidden beneath a canopy of jungle in the Madre de Dios region of Peru, it held thirty-one men and boys forced to work in illegal gold mining. They slept in makeshift shelters. They drank river water. They were beaten when they didn't meet quotas. To the outside world, they did not exist.

Then a satellite looked down.

High-resolution imagery from a commercial satellite detected what appeared to be new clearings in a protected forest area, clearings that matched the signatures of illegal mining operations. An AI system trained to identify forced labor indicators flagged the site for investigation. Three days later, a drone equipped with thermal imaging sensors flew over the camp at night, confirming the presence of dozens of heat signatures in structures too small and too crude to be voluntary housing.

The rescue operation was launched the following week. Thirty-one people walked out of that jungle. Many of them had been there for over a year.

From space to sky to ground. That is the new chain of liberation.

Eyes in the Sky: Drone Capabilities in Anti-Trafficking

Drones, unmanned aerial vehicles (UAVs), have evolved from military technology to versatile tools used across industries, from agriculture to disaster response. In anti-trafficking operations, drones offer capabilities that fundamentally change what is possible in surveillance, search, and rescue.

Thermal and night-vision imaging. Trafficking operations frequently occur in darkness and in locations designed to be invisible. Drones equipped with thermal cameras can detect human heat signatures through walls, vegetation, and darkness, revealing the presence of victims in locations that ground-level searches would miss. In maritime trafficking, where victims are transported in shipping containers or small boats, thermal drones can scan vessels quickly and safely.

Remote area coverage. Many trafficking operations deliberately locate in places that are geographically difficult to reach: dense forests, mountainous regions, remote islands, and vast agricultural estates. Drones can cover these areas in a fraction of the time required for ground teams, providing real-time video feeds that guide rescue planning and execution.

Evidence collection. Drones capture high-resolution video and imagery that serves as crucial evidence in trafficking prosecutions. Aerial footage of labor camps, fishing operations using forced labor, or illegal mining sites provides irrefutable documentation that strengthens legal cases and supports victim testimony.

Aid delivery. In situations where immediate rescue is not possible, drones can deliver essential supplies, food, water, medical kits, and communication devices, to victims awaiting extraction. This capability can be life-saving in remote locations where traditional supply chains cannot reach quickly enough.

Satellite Technology: The View from Space

While drones provide detailed, ground-level intelligence, satellites offer something equally vital: the big picture.

High-resolution monitoring. Modern commercial satellites, including those operated by Planet Labs, Maxar Technologies, and the European Space Agency's Copernicus program, produce imagery detailed enough to identify individual structures, vehicles, and even people from orbit. This capability enables monitoring of vast areas that would be impossible to cover with ground-based surveillance.

Change detection. One of the most powerful applications of satellite technology is change detection: comparing images of the same location over time to identify new construction, cleared land, or temporary structures that may indicate trafficking activity. A new compound appearing in a remote area, a sudden expansion of an agricultural operation, or the construction of makeshift housing near a border crossing, these changes are invisible from the ground but obvious from space.

Maritime surveillance. The fishing industry is one of the world's largest employers of forced labor. Satellite monitoring systems, including the Global Fishing Watch platform, track vessel movements worldwide, identifying ships that exhibit trafficking indicators: vessels that go "dark" by disabling tracking transponders, ships that transfer cargo or personnel at sea, and boats that operate in waters associated with forced labor practices.

The Environmental Justice Foundation has used satellite data to identify fishing vessels in Southeast Asian waters that employed enslaved crews. By tracking vessel movements and correlating them with port records and crew documentation, investigators were able to build cases that led to the liberation of hundreds of fishermen.

The Combined Force: Satellites Guide, Drones Execute

The most effective rescue operations integrate both technologies into a coordinated intelligence chain.

Satellite identifies the target. High-resolution imagery detects anomalies, new construction, unusual activity patterns, environmental changes, and flags potential trafficking sites.

Drones confirm and detail. UAVs deploy to flagged locations, providing close-range visual, thermal, and infrared data that confirms the presence of victims and maps the operational layout.

Ground teams execute. Armed with satellite context and drone-collected intelligence, rescue teams enter with detailed knowledge of the site's layout, the number of people present, guard positions, and access routes.

This integrated approach has been deployed against forced labor operations in brick kilns across South Asia, where satellite change detection identifies new kiln construction in regions known for bonded labor and drones confirm the presence of families living in conditions consistent with exploitation. Coordinated ground operations guided by this intelligence chain have freed dozens of individuals, including children, from conditions that would have remained invisible without the view from above.

Challenges and Ethical Boundaries

Drone and satellite technology are not without their complications.

Privacy and overreach. The ability to surveil from space and sky raises serious concerns about privacy and proportionality. Clear legal frameworks must govern when and how these technologies are deployed, ensuring they target trafficking operations without becoming tools of mass surveillance.

Resource disparities. High-quality satellite imagery and advanced drone operations require significant financial and technical resources. Developing countries, often those most affected by trafficking, may lack the infrastructure to deploy these tools effectively. International partnerships and technology-sharing agreements are essential to bridging this gap.

Operational security. Trafficking networks are adaptive. Once they become aware that drones or satellites are being used against them, they modify their operations: moving to covered locations, timing activities to avoid satellite passes, or using electronic countermeasures against drones. Continuous technological evolution is necessary to maintain operational advantage.

Thirty-one men and boys walked out of a jungle in Peru because a satellite saw what no human eye could see and a drone confirmed what no ground team could have reached in time.

The advantage is altitude: a broader field of vision, earlier warning, and fewer places for traffickers to disappear.

From that vantage, remoteness is less of a shield.

And from up there, no camp is hidden, no boat is invisible, no victim is beyond reach.

In the next chapter, we come back to earth, and examine the partnerships between technology companies and NGOs that are building the tools and infrastructure that make all of this possible.

Questions for Reflection and Discussion

1. Drone and satellite surveillance can locate hidden victims, but they also raise serious privacy concerns. How should governments regulate their use in anti-trafficking operations?

2. Many countries most affected by trafficking lack the resources for advanced drone and satellite programs. How can international partnerships bridge this technology gap?

3. What are the risks if traffickers learn to evade satellite detection, and how can the technology stay ahead?

4. How could maritime surveillance technology like Global Fishing Watch be expanded to cover other industries that rely on forced labor?

Chapter 8: Collaboration Between Tech Companies and NGOs

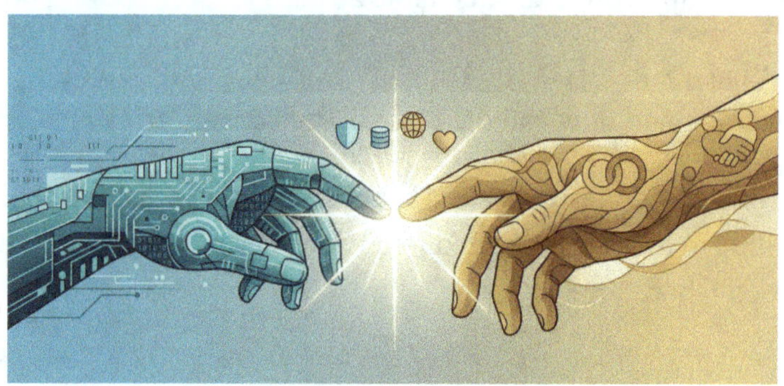

Where circuit meets compassion

When Ashton Kutcher and Demi Moore founded Thorn in 2012, the anti-trafficking world was skeptical. A Hollywood couple starting a tech nonprofit to fight child exploitation? It sounded like celebrity philanthropy at its most performative.

Over time, Thorn's work on tools such as Spotlight helped demonstrate that technology partnerships could move beyond awareness campaigns and into operational support for investigators. The broader point matters more than any single metric: when technical teams build with investigators and survivor advocates instead of at a distance from them, the resulting tools are far more likely to be used and trusted.

The skeptics were wrong. And the lesson is important: the fight against trafficking cannot be won by any single sector alone. It requires the collision of technological capability with frontline expertise. It requires the people who build the tools to sit in the same room as the people who know where the bodies are buried.

This chapter is about those rooms. And what gets built inside them.

Why Neither Side Can Do It Alone

The anti-trafficking movement has passion, expertise, and moral authority. What it often lacks is technological infrastructure, engineering talent, and the financial resources to build tools at scale.

The technology industry has exactly those resources. What it often lacks is the contextual understanding of how trafficking operates, what survivors need, and what frontline workers encounter daily.

When these two worlds come together, when engineers sit with social workers, when data scientists listen to survivors, when product managers collaborate with investigators, the results are transformative.

The Partnerships Changing the Game

Thorn and the Tech Coalition. Thorn's work represents perhaps the most successful tech-NGO partnership in anti-trafficking history. Beyond Spotlight, Thorn co-founded the Tech Coalition, an alliance of technology companies, including Apple, Google, Meta, Microsoft, Twitter, Snap, and others, committed to preventing online child exploitation. The Coalition's Project Protect develops shared technological infrastructure for detecting and removing exploitation content, creating efficiencies that no single company could achieve alone.

Microsoft and the International Centre for Missing & Exploited Children. Microsoft's PhotoDNA technology, which creates unique digital fingerprints of known CSAM images, is now used by over two hundred companies and NGOs worldwide. The technology has been donated to organizations free of charge, enabling even small nonprofits to deploy enterprise-grade image detection capabilities.

Google.org and anti-trafficking research. Google's philanthropic arm has provided millions of dollars in grants to anti-trafficking organizations, funded AI research for trafficking detection, and offered pro-bono engineering support to NGOs building technological tools. Google's support has been instrumental in developing data analytics platforms used by Polaris Project, the Global Modern Slavery Directory, and other organizations.

Amazon Web Services and the International Justice Mission. AWS provides cloud computing infrastructure and AI analytics tools to IJM, one of the world's largest anti-trafficking organizations. This partnership has accelerated IJM's investigative capabilities, enabling faster case processing, more sophisticated data analysis, and improved coordination across global offices.

Palantir and the National Center for Missing & Exploited Children. Palantir Technologies has partnered with NCMEC to deploy data integration and analysis tools that help investigators connect disparate data sources, tips, images, law enforcement reports, and CyberTipline submissions, into a unified intelligence picture.

Salesforce and anti-trafficking case management. Salesforce has provided its platform to multiple anti-trafficking organizations for case management, enabling nonprofits to track victim services, coordinate across agencies, and generate data that supports advocacy and policy development.

Building Bridges: How Effective Partnerships Work

Not every tech-NGO partnership succeeds. Many fail because of cultural differences, misaligned expectations, or the tech industry's tendency to build solutions in search of problems rather than listening first.

The partnerships that work share common elements:

They start with the problem, not the technology. The best partnerships begin with NGOs articulating what they need, not tech companies showing what they've built. Thorn's success was rooted in spending years embedded with law enforcement before writing a single line of code.

They build for the user. Anti-trafficking tools must be designed for the people who actually use them: investigators, social workers, hotline operators, and survivors. Elegance means nothing if the tool doesn't fit the workflow.

They commit for the long term. Trafficking is not a problem that can be solved with a six-month hackathon project. Effective partnerships require multi-year commitments to development, iteration, and support.

They center survivor voice. The most ethical and effective tools are designed with input from survivors themselves. Survivor advisory boards, trauma-informed design practices, and ongoing feedback loops ensure that technology empowers rather than retraumatizes.

The Funding Ecosystem

Money matters. And the way anti-trafficking technology is funded shapes what gets built.

Corporate social responsibility. Many tech companies fund anti-trafficking work through CSR programs, providing grants, engineering time, and infrastructure. While valuable, these contributions can be inconsistent, tied to corporate priorities that shift with market conditions.

Government funding. U.S. agencies including the Department of Justice, the Department of State's Office to Monitor and Combat Trafficking in Persons, and the Department of Homeland Security fund technology development for anti-trafficking purposes. The European Commission and UK Modern Slavery Innovation Fund are significant international funders.

Philanthropic investment. Foundations including the Minderoo Foundation (Walk Free), the Oak Foundation, and the Freedom Fund provide sustained funding for anti-trafficking technology development and deployment.

The gap remains significant. The global anti-trafficking technology sector is vastly underfunded relative to the scale of the problem. A single major tech company's annual R&D budget dwarfs the entire global investment in anti-trafficking technology. Closing that gap is not just a funding challenge, it is a moral imperative.

Open-Source Intelligence: How Anyone Can Contribute

You do not need to be a law enforcement officer or a data scientist to contribute to anti-trafficking intelligence. Open-source intelligence, commonly known as OSINT, refers to information gathered from publicly available sources, and it has become a powerful tool in the hands of trained advocates, journalists, and citizen investigators.

What OSINT is. OSINT encompasses any information that can be collected from public sources: social media profiles, online forums, business registries, domain registration records, satellite imagery, news reports, and government databases. When these disparate pieces of public information are assembled skillfully, they can reveal connections, locations, and patterns that would otherwise remain hidden.

Tools available to the public. Many OSINT tools are free and require no technical training. Reverse image search (Google Images, TinEye) can determine if a photo has been stolen or used across multiple fraud operations. Google dorking, which uses advanced search operators, can surface information that is technically public but not easily discoverable. Social media analysis tools like Maltego and Spiderfoot map relationships between accounts. The Wayback Machine archives historical web content that may have been deleted. And satellite imagery from Google Earth or Sentinel Hub can identify changes in land use or the construction of suspicious facilities.

Ethical boundaries. OSINT must be conducted within clear ethical and legal lines. Never attempt to contact a suspected trafficker or victim. Never hack, access private accounts, or misrepresent your identity. Never share unverified information publicly. Always report findings to law enforcement or established anti-trafficking organizations rather than acting independently. The goal of citizen OSINT is to gather and provide leads, not to conduct investigations.

Real-world impact. Citizen investigators have contributed to trafficking cases around the world. The Bellingcat investigative collective, while focused primarily on conflict and human rights, has pioneered OSINT methodologies that anti-trafficking advocates have adapted. TraffickCam, discussed in a previous chapter, is essentially a crowdsourced OSINT tool. And organizations like STOP THE TRAFFIK actively solicit community-sourced intelligence through their reporting platforms.

If you are reading this book and wondering how you can contribute without building software or joining law enforcement, OSINT may be your answer. The information is public. The tools are free. What is needed is care, discipline, and the willingness to look closely at what most people scroll past.

The most powerful technology in the world is useless without the human infrastructure to deploy it. And the most dedicated activists in the world are outmatched without the tools to fight at scale.

This is not about charity.
It is about alliance.
The code and the cause must be written together.

Questions for Reflection and Discussion

1. What would it take for a technology company you admire to make anti-trafficking a core part of its mission rather than a philanthropic side project?

2. How can OSINT tools be made more accessible to everyday advocates while maintaining ethical safeguards against misuse?

3. What risks arise when NGOs become dependent on corporate technology partners, and how can those risks be managed?

4. If you could design one technology tool to strengthen the collaboration between tech companies and anti-trafficking organizations, what would it do?

In the next chapter, we confront the most difficult questions of all: the legal and ethical boundaries that must govern how these powerful technologies are used, and what happens when they are not.

Chapter 9: Legal and Ethical Considerations in Using AI to Combat Trafficking

The scales of justice and innovation

Rosa was a domestic worker in the Gulf region. After she was rescued from a household where she had been exploited for two years, her face was scanned by a facial recognition system as part of the identification process. Her image was stored in a law enforcement database. Her story was documented in a case management system shared across multiple agencies.

Six months later, Rosa applied for asylum in a European country. During her interview, an immigration official pulled up her file, including photographs, her psychological evaluation, and details of her exploitation, that she had never consented to share beyond the rescue team. The information had been accessed through a data-sharing agreement she didn't know existed.

Rosa was not re-trafficked. She was re-traumatized. By the very systems designed to protect her.

This is the paradox at the heart of this chapter: the same technologies that save lives can, without proper safeguards, violate the dignity and autonomy of the people they're meant to serve. And if we do not reckon with that paradox honestly, we risk building a future where victims are rescued by systems they cannot trust.

The Right to Privacy in the Age of Surveillance

Every AI system discussed in this book, pattern recognition, facial recognition, financial monitoring, social media analysis, drone surveillance, runs on data. Vast quantities of data, collected from individuals who may or may not know they are being watched.

The ethical tension is real: aggressive data collection increases the chance of detecting trafficking. It also increases the risk of surveilling innocent people, exposing vulnerable populations, and creating databases that can be misused.

International data protection frameworks provide essential guardrails:

The EU General Data Protection Regulation (GDPR) establishes the principle of data minimization, collecting only the data necessary for a specific, stated purpose. It grants individuals the right to know what data is held about them, the right to access it, and the right to request its deletion. Anti-trafficking systems operating in EU jurisdictions must comply with these requirements, even when the purpose is law enforcement.

The California Consumer Privacy Act (CCPA) and similar state-level legislation in the United States provide consumers with rights to access and delete personal data, though law enforcement exceptions apply.

The UN Guiding Principles on Business and Human Rights establish the responsibility of technology companies to conduct human rights due diligence when their products are used in contexts that affect vulnerable populations.

But legal compliance is the floor, not the ceiling. Truly ethical anti-trafficking technology goes beyond what the law requires, building privacy protections into the architecture of the tools themselves, not just the policies that govern their use.

Algorithmic Bias: When AI Discriminates

AI systems learn from data. If that data reflects historical biases, and it almost always does, the AI will reproduce and amplify those biases at scale.

In anti-trafficking applications, the consequences can be devastating. Facial recognition systems that are less accurate for women and people of color may fail to identify victims from these demographics, or may generate false positives that waste investigative resources and subject innocent people to wrongful scrutiny.

Predictive models trained predominantly on sex trafficking data may be blind to labor trafficking patterns. Systems trained on data from wealthy countries may miss the indicators relevant to trafficking in the Global South.

Addressing algorithmic bias requires continuous, deliberate effort:

Diverse, representative training data. AI systems must be trained on datasets that reflect the full demographic, geographic, and typological diversity of trafficking. This requires intentional data collection and curation, not just the convenience of whatever data is available.

Regular independent auditing. Third-party audits of AI systems used in anti-trafficking work should assess accuracy across demographics, geographic regions, and trafficking types. Results should be published transparently.

Human oversight in all decisions. AI should inform, not determine. No arrest, no rescue operation, and no victim identification should be based solely on algorithmic output. Human investigators must validate AI-generated intelligence before acting on it.

Survivor Consent and Data Autonomy

Survivors of trafficking are not data points. They are people who have had their autonomy stolen, and the systems designed to help them must not steal it again.

Informed consent. Survivors must be clearly informed about what data is collected, how it will be used, who will have access, and for how long. Consent must be voluntary, specific, and revocable. A survivor who agrees to share her story for a VR training tool must have the right to withdraw that consent at any time.

Anonymization as default. Victim data should be anonymized by default, with identifying information accessible only to authorized personnel for specific, justified purposes.

Survivor-centered design. Anti-trafficking technology should be designed with input from survivors. Survivor advisory boards should review tools before deployment, assessing not just functionality but potential for harm, retraumatization, or disempowerment.

Transparency, Accountability, and the Black Box Problem

Many AI systems operate as "black boxes", their decision-making processes are too complex for humans to understand or explain. In medical diagnosis or product recommendation, opacity is inconvenient. In law enforcement and anti-trafficking work, it is dangerous.

If an AI system flags someone as a suspected trafficker, the reasoning must be explainable. If facial recognition identifies a person as a missing victim, the confidence level and methodology must be transparent. Courts require it. Ethics demand it.

Explainable AI (XAI) is an emerging field dedicated to making algorithmic decision-making transparent and interpretable. Anti-trafficking applications should adopt XAI principles as standard practice, ensuring that every alert, every flag, and every identification can be understood, questioned, and audited.

Independent oversight bodies. AI tools deployed in anti-trafficking work should be subject to review by independent ethics committees that include technologists, legal experts, human rights advocates, and survivors. These bodies should have the authority to recommend modifications, restrictions, or suspension of tools that fail to meet ethical standards.

International Law and Cross-Border Complexity

Human trafficking is inherently transnational. Victims cross borders. Data crosses borders. Investigations cross borders. But legal frameworks do not always follow.

Collecting data on a trafficker in one country, analyzing it with servers in another, and sharing intelligence with law enforcement in a third creates a jurisdictional web that existing legal frameworks struggle to navigate. The Budapest Convention on Cybercrime, the Palermo Protocol on Trafficking, and various bilateral mutual legal assistance treaties provide some framework, but gaps remain.

The EU's AI Act, which entered into force in August 2024 with phased implementation extending through 2026, represents the most comprehensive attempt to regulate AI across borders, classifying anti-trafficking surveillance applications as high-risk and imposing transparency, accuracy, and human oversight requirements. Whether other jurisdictions will adopt similar frameworks remains to be seen.

The Dual-Use Danger

Every tool in this book can be misused.

Facial recognition designed to find trafficking victims can be deployed to surveil ethnic minorities. Predictive analytics built to anticipate trafficking hotspots can be repurposed for discriminatory policing. Drone technology intended for rescue operations can become instruments of oppression in authoritarian states.

This dual-use risk is not hypothetical. It is documented. And it demands that the anti-trafficking community builds safeguards into the technology itself, not just the policies governing its use.

Strict licensing and access controls for anti-trafficking AI tools, ensuring they are available only to vetted, trained, and accountable organizations.

Built-in audit trails that log every use of the technology, creating accountability for how and when tools are deployed.

Human rights impact assessments before deployment in any new jurisdiction or context, evaluating the risk of misuse against the potential for benefit.

Rosa eventually received asylum. Her case was reviewed, the data-sharing breach was documented, and the agency responsible implemented new consent protocols. It should not have taken her retraumatization to prompt the change.

Ethics are not optional.
They are not the fine print.
They are the foundation.

In the final chapter, we look forward, imagining a future where these technologies, governed by these principles, converge to create a world where trafficking is not just fought but rendered obsolete.

Questions for Reflection and Discussion

1. How can anti-trafficking organizations ensure that the technology they deploy respects survivor autonomy and consent, not just in policy but in practice?

2. Should there be an international treaty governing the use of AI in anti-trafficking work? What should it include?

3. How do you weigh the potential for AI surveillance to save lives against the risk that it could be repurposed for political oppression?

4. What does "explainable AI" look like in practice, and why is it essential for anti-trafficking applications that involve law enforcement?

Chapter 10: The Future of Technology in the Fight Against Human Trafficking

From present tools to future light

Picture this: the year is 2035. A seventeen-year-old named Sana in a refugee camp in East Africa receives a notification on her phone. It is a job offer from a "recruitment agency" promising domestic work in a wealthy Gulf state. The salary is unusually high. The description is vague.

Before Sana can respond, a warning appears on her screen. An AI system integrated into the camp's digital infrastructure has analyzed the message in real time, cross-referencing it against a global database of known trafficking recruitment patterns. The sender's phone number is linked to a network flagged by blockchain-verified intelligence from Interpol. The "agency's" website was registered three days ago. The language patterns match templates used in documented trafficking operations across the region.

The notification on Sana's screen does not say "scam." It says: "This message has been flagged as a potential trafficking recruitment attempt. If you believe you are in danger, press here. A counselor is available now."

Sana never gets on the plane. The recruitment network is flagged for investigation. The system that protected her was built by the technologies we are developing today.

This is not science fiction. Every component of that scenario, the AI analysis, the blockchain intelligence, the real-time warning, the integrated response, exists in some form right now. The future is not about inventing new magic. It is about connecting what we already have into systems powerful enough to outpace the traffickers.

Quantum Computing: Processing at the Speed of Justice

Quantum computing represents a fundamental leap in computational power: the ability to process complex calculations exponentially faster than any classical computer. For anti-trafficking work, the implications are profound.

Cracking encrypted criminal communications. Trafficking networks increasingly use end-to-end encryption to shield their operations. Quantum computing could potentially decrypt these communications in minutes rather than years, opening a window into criminal planning and coordination that is currently opaque.

Real-time multi-source intelligence fusion. Today, synthesizing data from surveillance cameras, financial records, social media, satellite imagery, and law enforcement databases takes time, time that traffickers use to move victims. Quantum processors could integrate all of these data streams simultaneously, creating a comprehensive, real-time picture of trafficking activity across an entire region.

The challenge is dual-use: as quantum computing advances, traffickers will also gain access to more powerful tools. The anti-trafficking community must stay ahead, investing in post-quantum cryptography to protect victim data while developing quantum-enhanced analytics for detection and disruption.

The Internet of Things: A Network of Sentinels

The Internet of Things (IoT) is a vast and growing network of interconnected devices, sensors, cameras, trackers, and smart systems, that collect and share data continuously. As IoT infrastructure expands, its potential for anti-trafficking applications grows.

Smart city integration. Traffic sensors, license plate readers, public surveillance cameras, and transit monitoring systems can feed into centralized AI platforms that detect trafficking-related movement patterns. A vehicle that frequently visits locations associated with commercial sexual exploitation. A pattern of short-term hotel stays by the same individual with different companions. Transit card data showing a minor traveling repeatedly between cities during school hours.

Wearable safety devices. For populations at heightened risk, survivors in recovery, migrant workers, unaccompanied minors, discreet wearable devices could provide a lifeline. A bracelet or pendant with GPS tracking and a panic button, linked to a response network, could signal for help if a re-trafficking attempt occurs. Several humanitarian organizations and technology partners have begun piloting such devices in vulnerable communities.

Smart supply chain monitoring. IoT sensors embedded throughout supply chains, in factories, on shipping containers, in warehouses, can continuously monitor working conditions, recording temperature, noise levels, movement patterns, and hours of operation. Anomalies that suggest forced labor conditions, workers present at 3 a.m. seven days a week, for instance, can trigger automatic alerts.

Multi-Layered Predictive Platforms: Forecasting Exploitation

The future of anti-trafficking technology lies not in any single tool but in the integration of multiple technologies into unified predictive platforms.

Trafficking prediction engines. Imagine a platform that integrates economic data (unemployment rates, currency crises), climate data (droughts, floods, natural disasters), conflict monitoring (civil unrest, military operations), migration tracking (refugee movements, visa overstays), and social media analysis (recruitment language, distress signals) into a single predictive model. Such a platform could identify populations at elevated risk of trafficking weeks or months before traffickers reach them, enabling preventive interventions rather than reactive rescues.

Simulation and scenario planning. Predictive platforms could model the trafficking impact of anticipated events: how will the displacement caused by an approaching hurricane increase trafficking risk in coastal communities? How will the economic fallout of a trade embargo create vulnerability in manufacturing regions? These simulations would enable governments and NGOs to pre-position resources and protections.

AI-Generated Decoys and Digital Counter-Trafficking

As AI becomes more sophisticated, so do its applications in active counter-trafficking operations.

Digital decoy operations. AI-generated personas, complete with social media profiles, communication histories, and realistic interaction patterns, can be deployed on dark web forums and illicit platforms to bait traffickers. These digital decoys engage suspects in conversations that reveal networks, methods, and planned operations, gathering evidence for prosecution without putting any human operative at risk.

Counter-propaganda and interception. Traffickers use digital platforms to recruit victims. Future AI systems could intercept these recruitment attempts in real time, inserting counter-messages, redirecting vulnerable individuals to support services, and disrupting grooming conversations before exploitation occurs.

A Decentralized Global Registry

Blockchain and Web3 technologies are paving the way for a decentralized global trafficking intelligence system: a shared ledger governed by a consortium of vetted law enforcement agencies, NGOs, and survivor-led organizations.

Immutable case records. Every trafficking case, from initial report through investigation, prosecution, and survivor recovery, could be recorded on a decentralized blockchain, creating a continuous, tamper-proof record that supports accountability, prevents duplication, and ensures continuity of care across borders.

Cross-border evidence sharing. Cryptographically verified evidence stored on blockchain could be accessed by authorized investigators in any jurisdiction, eliminating the delays and bureaucratic barriers that currently hamper international cooperation.

Survivor-Led Technology Design

The most important technological future is not about machines. It is about who shapes them.

Human-centered AI. The next generation of anti-trafficking tools must be designed with survivors at the table, not as consultants brought in after the product is built, but as co-designers who shape the tool from conception through deployment. Survivor-informed algorithms are more accurate, more ethical, and more trusted by the communities they serve.

Digital empowerment. Survivors equipped with digital literacy, cybersecurity training, and technical skills can become active participants in the anti-trafficking ecosystem, monitoring online spaces, reporting leads, designing tools, and serving as paid consultants to law enforcement and technology companies. This is not charity. It is recognition that survivors possess expertise that no algorithm can replicate.

A Unified Global Strategy

Technology alone cannot end trafficking. It requires global alignment: shared standards, interoperable systems, and cooperative governance.

Tech-enabled multinational task forces. International coalitions equipped with shared platforms, integrating AI, blockchain, satellite, and Big Data capabilities, could enable real-time, coordinated responses to trafficking operations that span multiple countries.

Universal ethical standards. Global agreements governing the ethical use of AI in anti-trafficking work, building on the EU's AI Act, the OECD AI Principles, and the UNESCO Recommendation on AI Ethics, could establish a baseline of accountability, transparency, and human rights protection.

Sana is not a real person. But she represents the future we are building: a future where the vulnerable are protected before exploitation begins, where intelligence flows faster than trafficking networks can adapt, and where technology serves as a shield, not a weapon.

The future is not determined.
It is designed.
And we are the designers.

But as we close this final chapter on technology, one question remains, perhaps the most important question in this entire book. It is not about code or satellites or blockchain. It is about something deeper. Something no algorithm can solve.

Turn the page.

Questions for Reflection and Discussion

1. Which emerging technology discussed in this chapter do you believe has the greatest potential to reduce trafficking in the next decade, and why?

2. How can the anti-trafficking movement ensure that survivors are co-designers of the technology built to protect them, not just beneficiaries?

3. What would a world where trafficking is "technologically impossible" actually look like? What systems would need to be in place?

4. After reading this book, what is one concrete action you can take in the next thirty days to contribute to the fight against trafficking?

Chapter 11: The Trafficker's Tech Playbook

The dark chessboard of exploitation

We have spent ten chapters discussing the technologies being built to fight trafficking. Now we need to talk about the technologies being built to enable it.

One reality the anti-trafficking field cannot afford to romanticize is this: many traffickers learn quickly, test new methods, and adopt digital tools as soon as those tools improve reach, secrecy, or profit. They study platform features, exploit automation, and adjust tactics with the speed of a startup, except their business model is coercion.

If we do not understand how traffickers use technology, we cannot build systems to stop them. You do not defeat an enemy you refuse to study.

The Startup That Sells People

Modern trafficking operations mirror legitimate technology companies in structure, strategy, and speed. A mid-tier trafficking network might include recruiters who function as marketing teams, identifying and targeting vulnerable populations through social media campaigns. Logistics coordinators manage transportation, housing, and document fraud with the precision of a supply chain operation. Financial controllers handle revenue through layered cryptocurrency wallets, prepaid cards, and shell companies. And at the top, operators treat the enterprise as a portfolio, diversifying across sex trafficking, labor trafficking, and organ trafficking to manage risk.

This is not a street corner operation. This is organized crime running on the same digital infrastructure that powers legitimate businesses. They use project management tools. They use encrypted group chats. They use analytics to track which recruitment messages convert and which do not.

Understanding their playbook is the first step to dismantling it.

Deepfakes as Weapons of Control

Generative AI has given traffickers a new and terrifying tool: the ability to create synthetic images and videos of their victims that never actually occurred. Deepfake technology, which uses neural networks to superimpose one person's likeness onto another's body, is being weaponized in multiple ways.

Sextortion and blackmail. Traffickers create fabricated explicit images of victims and threaten to distribute them to the victim's family, school, or community unless the victim complies with demands. The images do not need to be real to be devastating. The shame and fear they produce are real, and they are extraordinarily effective at maintaining control.

Fraudulent recruitment. Deepfake videos of fake employers, fake agency representatives, or fake family members are used to build trust with potential victims. A video call with a "hiring manager" who appears professional and legitimate can be entirely AI-generated. The victim believes they are speaking to a real person. They are speaking to a script.

Identity fraud. Traffickers use AI face-swapping technology to create fraudulent identity documents, bypass video verification systems, and open financial accounts under fabricated identities. This allows them to move money, rent properties, and book travel without leaving a traceable identity behind.

The counter-trafficking response is developing detection tools, including deepfake classifiers that analyze pixel patterns, facial inconsistencies, and audio artifacts that betray synthetic content. But the arms race is fierce, and the generation speed of deepfakes far outpaces the detection capacity of current tools.

Algorithm-Driven Recruitment

Social media algorithms are designed to connect people with content they are likely to engage with. Traffickers have learned to exploit this design with surgical precision.

Targeted advertising of fraudulent opportunities. Traffickers create polished social media posts advertising jobs, modeling opportunities, romantic relationships, or travel experiences. They target these posts at demographics that data analysis has identified as vulnerable: young women in economically depressed regions, LGBTQ+ youth who have been rejected by their families, migrants seeking work in foreign countries, and teenagers with public profiles showing signs of emotional distress.

Grooming through engagement algorithms. When a potential victim engages with a recruiter's content, liking a post, responding to a comment, clicking a link, the platform's algorithm amplifies the connection. It suggests more content from the recruiter. It surfaces the recruiter in the victim's feed. It facilitates a deepening relationship that the victim believes is organic but that the recruiter has engineered.

Burner accounts at scale. Traffickers use automation tools to create and manage dozens or hundreds of social media accounts simultaneously. When one account is flagged or banned, another is already active. They use VPNs, virtual phone numbers, and disposable email addresses to create accounts that are difficult to trace. Some operations use AI chatbots to handle initial outreach, only bringing in human operators once a potential victim has shown interest.

Social media companies have made progress in detecting and removing trafficking-related content, but the volume is overwhelming. Meta alone reported removing millions of pieces of content related to human exploitation in 2024. The problem is not awareness. The problem is scale.

The Encrypted Infrastructure

Encryption is a fundamental right. It protects journalists, activists, and ordinary citizens from surveillance and oppression. But it also protects traffickers.

End-to-end encrypted messaging apps like Signal, Telegram, and WhatsApp are used extensively by trafficking networks for operational communication. Group chats coordinate logistics. Voice messages deliver instructions that leave no text record. Disappearing messages erase evidence automatically. Encrypted file sharing distributes exploitation material without ever touching a public server.

Traffickers also use encrypted email services, virtual private networks, and the Tor browser to conduct business on the dark web. They rotate phone numbers using eSIM technology, sometimes changing numbers daily. They use cryptocurrency mixing services to launder payments. And they train their operatives in operational security, teaching them to compartmentalize information so that the arrest of one member does not compromise the entire network.

The policy debate around encryption and law enforcement access is complex and ongoing. What is clear is that the anti-trafficking community needs tools that can detect trafficking patterns without breaking encryption, using metadata analysis, behavioral signals, and network mapping rather than content interception.

The Disposable Digital Identity

One of the most effective tactics in the trafficker's playbook is the creation of disposable digital identities that can be burned and rebuilt in minutes.

A trafficker operating today might use a prepaid phone purchased with cash, a VPN routing traffic through three countries, a Telegram account registered with a virtual phone number, a cryptocurrency wallet created anonymously, and a social media profile using AI-generated profile photos that depict a person who does not exist. All of this can be set up in under an hour and destroyed in seconds.

This digital disposability is the trafficker's greatest advantage. It means that traditional investigative methods, which depend on persistent identities, traceable accounts, and stable digital footprints, are often chasing ghosts.

The response requires a shift from identity-based investigation to pattern-based investigation: tracking behavioral signatures, communication rhythms, financial flow patterns, and network structures that persist even when individual identities are discarded.

Understanding the trafficker's playbook is not meant to inspire fear. It is meant to inspire precision.

You cannot outbuild what you do not understand.
You cannot outrun what you refuse to study.
Know the enemy. Then build the tools that make their playbook obsolete.

Questions for Reflection and Discussion

1. How does understanding the trafficker's use of technology change the way you think about everyday platforms like social media, messaging apps, and cryptocurrency?

2. What responsibility do social media companies bear for the exploitation that occurs on their platforms, even when traffickers circumvent content moderation?

3. How can the anti-trafficking community develop tools that detect trafficking patterns without undermining the encryption that protects legitimate users?

4. If traffickers operate like tech startups, what lessons from the startup world could anti-trafficking organizations adopt to match their speed and adaptability?

Chapter 12: The Survivor Technology Toolkit

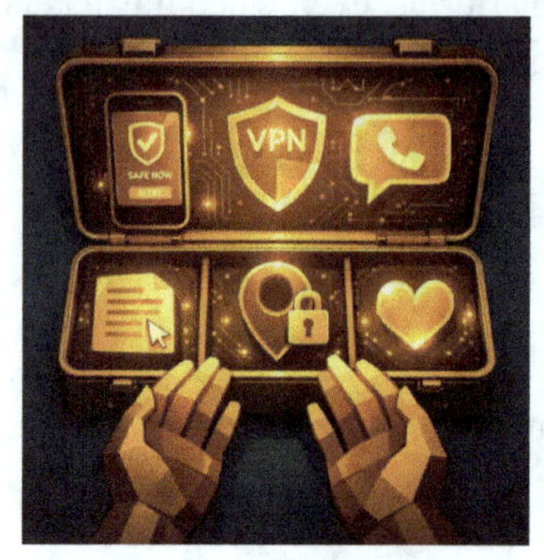

Tools of empowerment, not surveillance

This chapter is different from every other chapter in this book.

It is not written about survivors. It is written for them. And for the advocates, social workers, shelter staff, and allies who walk beside them.

Most anti-trafficking books focus on the tools that law enforcement and technologists use to find victims and prosecute traffickers. Almost none address a critical gap: what happens after rescue? How does a survivor navigate a digital world that was used to exploit them? How do they protect themselves online? How do they rebuild a digital identity that is safe, empowering, and their own?

This chapter is a practical, plain-language guide to digital self-defense and digital rebuilding for trafficking survivors. It is not exhaustive, and technology changes fast. But the principles here are foundational, and they are designed to be accessible regardless of technical background.

Digital Safety After Rescue: The First Steps

When a survivor exits a trafficking situation, their digital life is often compromised. Traffickers may have access to their email, social media accounts, phone, and location data. The first priority is securing the digital perimeter.

Change all passwords immediately. Every account the survivor has ever used should be treated as compromised. Use a new email address as the recovery email for all accounts. Choose passwords that are long and unique. A password manager app (such as Bitwarden, which is free and open-source) can generate and store strong passwords so the survivor does not need to memorize them.

Enable two-factor authentication (2FA). This adds a second layer of security beyond a password, usually a code sent to a phone or generated by an app. Even if a trafficker has a password, they cannot access the account without the second factor. Prioritize 2FA on email, banking, and social media accounts.

Get a new phone number. If the trafficker knows the survivor's phone number, that number is a tracking vector. A new prepaid phone or a new number through the carrier should be obtained as soon as safely possible. The old number should not be ported or forwarded, as this can reveal the new number.

Review location sharing. Check all apps for location-sharing permissions. Google Maps, Find My Friends, social media apps, and even photo gallery apps can broadcast a survivor's location. Turn off location services for all apps except navigation, and only enable it when actively needed.

Scrubbing the Digital Footprint

Traffickers may have created online profiles, advertisements, or social media accounts using the survivor's real name, photos, or personal information. This digital footprint can follow a survivor for years, affecting employment, housing, relationships, and mental health.

Search for yourself online. Use Google, Bing, and DuckDuckGo to search the survivor's name, known aliases, phone numbers, and email addresses. Document what appears. This is the starting point for cleanup.

Request content removal. Most platforms have processes for removing non-consensual intimate images, fraudulent profiles, and content posted without consent. Google has a specific removal request process for involuntary explicit imagery. The Cyber Civil Rights Initiative (cybercivilrights.org) provides guides and support for content removal.

Use data removal services. Services like DeleteMe, Kanary, and Privacy Duck specialize in removing personal information from data broker websites that aggregate and sell personal data. Many anti-trafficking organizations can cover the cost of these services for survivors.

Deactivate, do not just delete. When closing old social media accounts, first download any personal photos or data the survivor wants to keep, then deactivate and formally delete the account. Simply deactivating without deleting leaves the data accessible.

Encrypted Communication: Speaking Safely

Survivors need communication tools that traffickers and abusers cannot intercept.

Use end-to-end encrypted messaging. Signal is widely regarded as the most secure free messaging app available. Messages, calls, and video chats are encrypted so that only the sender and recipient can read them. Not even Signal's own servers can access the content.

Use a VPN for internet browsing. A Virtual Private Network encrypts internet traffic and masks the user's location. Free options like ProtonVPN offer basic protection. A VPN prevents anyone on the same Wi-Fi network, including a trafficker or abuser, from monitoring the survivor's online activity.

Be cautious with email. Standard email (Gmail, Yahoo, Outlook) is not end-to-end encrypted by default. For sensitive communications, ProtonMail offers free encrypted email. For everyday use, enabling 2FA on existing email accounts is the minimum.

Understand disappearing messages. Signal and WhatsApp both offer disappearing messages that automatically delete after a set time. This can be useful for survivors who want to communicate without leaving a digital trail. However, screenshots can still be taken, so no digital communication is ever perfectly safe.

Apps and Platforms Built for Survivors

A growing ecosystem of apps and platforms has been designed specifically for survivor safety and empowerment.

Safety planning apps. Apps like myPlan help survivors create personalized safety plans, assess danger levels, and connect with local resources. These apps are designed with discretion in mind, often disguisable as other types of apps on the phone's home screen.

Financial rebuilding tools. Many survivors exit trafficking with no credit history, no bank account, and no financial literacy. Organizations like Justine Petersen and other community development financial institutions offer programs that help individuals with no credit history build credit, open safe bank accounts, and develop financial independence, services that are critical for trafficking survivors starting over.

Job training and digital literacy platforms. Programs like AnnieCannons train survivors in software development, creating a direct pathway from exploitation to technology careers. Survivors learn to code, build professional portfolios, and enter the workforce with marketable, well-paying skills.

Survivor networks and peer support. Online communities moderated by and for survivors provide connection, mentorship, and mutual support. Organizations like the National Survivor Network and Survivor Alliance maintain platforms where survivors can connect safely with peers who understand their experience.

A Digital Self-Defense Framework

Digital safety is not a one-time fix. It is an ongoing practice. Here is a framework survivors can return to regularly:

Audit. Every three months, search for your name and known information online. Check account security settings. Review app permissions on your phone. Look for anything that should not be there.

Protect. Keep passwords unique and strong. Keep 2FA enabled. Keep your VPN active when using public Wi-Fi. Keep your phone number private and share it selectively.

Build. Create new accounts that are yours alone. Build a professional online presence through LinkedIn or a personal website. Learn new digital skills. Take ownership of your digital identity as an act of reclamation.

Connect. Reach out to organizations that provide digital safety support for survivors. You do not have to do this alone. The National Human Trafficking Hotline (1-888-373-7888) can connect you to local resources, and the Cyber Civil Rights Initiative provides direct support for online exploitation.

Technology was used against you. Now it belongs to you.

Your passwords are your walls.
Your encrypted messages are your shield.
Your digital literacy is your power.
No one gets to take that from you again.

Questions for Reflection and Discussion

1. How can shelters and survivor service organizations integrate digital safety training into their existing programs?

2. What barriers might survivors face in accessing the digital tools described in this chapter, and how can those barriers be addressed?

3. How does digital empowerment contribute to long-term recovery and independence for trafficking survivors?

4. What role can technology professionals play in volunteering their skills to help survivors secure their digital lives?

An Open Letter to the Tech Industry

A letter that demands an answer

To the founders, the engineers, the product managers, the venture capitalists, the board members, and the interns. To everyone who builds, ships, scales, and iterates. To Silicon Valley and every tech hub on every continent. This letter is for you.

I am not writing this to shame you. I am writing this to call you in.

You build platforms that connect billions of people. You design algorithms that shape what the world sees, reads, believes, and desires. You control infrastructure that processes more transactions in a day than most governments handle in a year. You have, without exaggeration, more power to shape human experience than any industry in the history of civilization.

And human trafficking is running on your rails.

Victims are recruited on your social media platforms. They are advertised on your classified sites. They are sold through your encrypted messaging apps. Their exploitation is paid for with cryptocurrency that flows through your exchanges. Their abuse is distributed on networks your infrastructure supports. And the algorithms you designed to maximize engagement are, right now, connecting predators with prey.

I know you did not design these systems to enable trafficking. I know that the vast majority of people in technology are horrified by exploitation and would stop it if they knew how. That is why I am writing this: because you can. And because the gap between what you could do and what you are doing is the space where millions suffer.

What You Could Do Tomorrow

Fund it like you fund everything else. The entire global investment in anti-trafficking technology is a rounding error on any major tech company's R&D budget. A single company's annual spending on AI research exceeds the total funding available to every anti-trafficking tech initiative on earth combined. Dedicate one percent of your AI research budget to anti-trafficking applications. One percent. That alone would transform the field.

Build detection into the product, not the afterthought. Trafficking detection should not be a policy team's side project.

It should be embedded in product design from day one. When you build a messaging platform, build trafficking-pattern detection into the architecture. When you build a payment system, build financial anomaly detection for trafficking typologies. When you build a social platform, build grooming-detection algorithms into the engagement engine. Do not wait for legislation to force you. Lead.

Share your data. You have more data on human behavior than any institution in history. Anti-trafficking researchers are working with datasets that are a fraction of what you could provide, anonymized, aggregated, and ethically governed. Create data-sharing partnerships with vetted anti-trafficking organizations. The patterns are in your data. Let the people who know how to read them see it.

Hire survivors. Not as spokespeople. Not as token consultants. As engineers, designers, product managers, and advisors who bring lived experience to the teams building the tools. Survivors understand how trafficking operates online in ways that no amount of academic research can replicate. Their expertise is not charity. It is competitive advantage.

Slow down when the stakes are human. The "move fast and break things" ethos has produced extraordinary innovation. It has also produced platforms that were scaled to billions of users before anyone considered how they could be weaponized for exploitation. When you build tools that touch vulnerable populations, slow down. Conduct human rights impact assessments. Test for misuse scenarios before launch, not after the first lawsuit.

The Legacy Question

Every technology leader will be remembered for something. The question is whether you will be remembered for what you built or for what you allowed to happen on what you built.

History is not kind to those who had the power to act and chose to optimize quarterly earnings instead. History remembers the companies that used their reach to protect the vulnerable, to set standards, to prove that profit and principle are not mutually exclusive.

You have the engineers. You have the data. You have the infrastructure. You have the resources. The only thing missing is the decision.

Make it.

The world your platforms have connected is watching.

The survivors your systems failed are waiting.

And the next generation of builders is deciding right now what kind of industry they want to inherit.

Show them something worth inheriting.

Damien Brinson

Founder & Executive Director, Deebo Haven Center

Part III

Final Reflections and Action

Epilogue: Beyond Innovation, Toward a Culture of Uninterrupted Freedom

From broken chains, a tree of light

There will come a day when the phrase "human trafficking" will sound like a relic: an ancient term, once necessary, now nearly forgotten. A day when modern slavery will be regarded as not just illegal but technologically impossible, socially unimaginable, and culturally obsolete.

But that day isn't guaranteed. It has to be engineered.

And not just with better tools, but with better values embedded into the systems we build, the policies we enforce, and the everyday decisions we make as individuals, communities, and societies.

What Happens When Technology Works?

Imagine this: We succeed. AI disrupts trafficking rings before victims are moved. Satellites detect high-risk zones in real time. Blockchain verifies ethical labor across every product. VR trains entire nations to identify and respond to coercion. Survivors are not only rescued but elevated, educated, employed, and empowered.

What then?

The next frontier becomes human culture. Because no amount of code, drones, or predictive analytics can solve what's coded deep inside culture: the demand for exploitation. The economic dependence on underpaid labor. The willful blindness of comfortable societies.

These cannot be automated away. They must be confronted, with courage, consciousness, and an unwillingness to look the other way.

The Cultural Technology We Rarely Talk About

There is another kind of technology that shapes the future:

The technology of belief.
The operating system of a society.

Every movement, every policy, every breakthrough is downstream from what we believe to be normal, acceptable, or inevitable.

To end trafficking permanently, we must:

Build shame-resistant communities, where victims can safely come forward without fear of being disbelieved, blamed, or judged.

Create ethical consumer cultures, where people understand how exploitation hides in cheap products and fast fashion, and choose better.

Raise digitally moral generations, where the next wave of tech leaders don't just code apps but carry convictions.

Cultivate a culture of seeing, where we train our eyes not to look away when exploitation hides in plain sight.

We cannot algorithm our way out of moral apathy.
We must reprogram the social code, too.

Freedom Must Be Designed into the Future

Architects design buildings to withstand storms.
Coders design software to handle threats.
What if we designed cities, markets, schools, and governments to protect freedom as a default setting?

What if zoning laws ensured no neighborhood was isolated enough for trafficking to hide?

What if job platforms used blockchain to ensure recruiters are vetted before anyone ever applies?

What if schools taught digital consent and exploitation awareness by fifth grade, globally?

What if we required Ethical Innovation Certifications for all technology tools developed with social data?

This is the next phase of anti-trafficking work: not just fighting back, but building forward.

You Are a Link in the Chain: Or a Break in It

This book wasn't written for spectators.

It was written for system builders. Cultural engineers. Soul warriors.

Anyone bold enough to believe that we can end exploitation in our lifetime, if we design for it.

Whether you're a teacher, investor, coder, policymaker, parent, student, or survivor, there is space for you in this global counterforce.

You don't have to understand AI. You just have to understand agency.

Yours. Theirs. Ours.

Because freedom, like trafficking, spreads through systems. And it begins when someone decides to stand in the gap.

The Final Word

The opposite of trafficking isn't just rescue.
It's restoration.
It's people returning to the world not as damaged property
but as undeniable proof that light moves faster than darkness.

That future is still ours to create.

Not just with circuits and satellites,
but with solidarity, story, and sacred determination.

The next breakthrough is not only technological.
It's moral.
And it starts now, with you.

Damien Brinson

Founder & Executive Director, Deebo Haven Center

Orlando, Florida

Resource Appendix

Tools, Organizations, and Support for Ending Human Trafficking

Whether you are a survivor, advocate, policymaker, educator, technologist, or concerned citizen, the following resources can help you engage further in the fight against human trafficking. All URLs were verified at the time of publication.

[Audiobook Narrator Note: This appendix contains extensive web links and contact information. Listeners are encouraged to visit deebohavencenter.org/resources for a complete, clickable version of all resources listed below.]

Emergency Hotlines and Reporting Services

National Human Trafficking Hotline (U.S.)

Call: 1-888-373-7888 | Text: "BEFREE" (233733)

Website: https://humantraffickinghotline.org

U.S. Department of Homeland Security: Blue Campaign

Report suspicious activity or access educational materials.

Website: https://www.dhs.gov/blue-campaign

Polaris Project

Operates the U.S. trafficking hotline and provides data-driven resources.

Website: https://polarisproject.org

International Organization for Migration (IOM)

Cross-border assistance and migration-related exploitation support.

Website: https://www.iom.int/human-trafficking

UNODC Human Trafficking Resources (Global)

Country-specific reporting guides and global intelligence.

Website: https://www.unodc.org/unodc/en/human-trafficking

National Center for Missing & Exploited Children (NCMEC)

CyberTipline for reporting online child exploitation.

Website: https://www.missingkids.org

Nonprofit Organizations and Survivor Support

Deebo Haven Center (Orlando, FL)

A safe haven for survivors offering support, advocacy, and restoration.

Website: https://www.deebohavencenter.org

International Justice Mission (IJM)

Legal advocacy and rescue missions in over thirty countries.

Website: https://www.ijm.org

Love146

Prevention programs and long-term survivor care.

Website: https://www.love146.org

Thorn

Builds technology to defend children from online sexual exploitation.

Website: https://www.thorn.org

Free the Slaves

Works to eradicate modern slavery through systemic change and survivor leadership.

Website: https://www.freetheslaves.net

A21

Global anti-trafficking organization with prevention and aftercare programs.

Website: https://www.a21.org

Coalition to Abolish Slavery and Trafficking (CAST)

Legal and social services for survivors in Los Angeles.

Website: https://www.castla.org

Shared Hope International

Training, policy advocacy, and victim care for domestic minor sex trafficking.

Website: https://sharedhope.org

ECPAT International

Global network working to end the sexual exploitation of children.

Website: https://ecpat.org

Global Fund to End Modern Slavery (GFEMS)

Public-private partnership that invested in anti-slavery programs worldwide. GFEMS concluded operations in 2024; its research archive remains a valuable resource.

Website: https://www.gfems.org

Anti-Trafficking Technology Tools and Platforms

Spotlight by Thorn

AI-based tool used by law enforcement to find trafficking victims faster.

Website: https://www.thorn.org/spotlight

Chainalysis

Blockchain analytics platform for tracing cryptocurrency in criminal investigations.

Website: https://www.chainalysis.com

TraffickCam

Crowdsourced hotel image database used to identify trafficking locations.

Website: https://traffickcam.com

Global Fishing Watch

Satellite monitoring of fishing vessel activity to detect forced labor at sea.

Website: https://globalfishingwatch.org

Counter-Trafficking Data Collaborative (CTDC)

Global hub for trafficking data, maintained by IOM.

Website: https://www.ctdatacollaborative.org

Tech Against Trafficking (TAT)

Coalition of tech companies supporting anti-trafficking technology development.

Website: https://www.techagainsttrafficking.org

STOP THE TRAFFIK

Community-based prevention and data intelligence platform.

Website: https://www.stopthetraffik.org

Educational and Training Resources

SOAR Online Training (U.S. DHHS)

Free training for healthcare and social service professionals.

Website: https://www.acf.hhs.gov/otip/training/soar-to-health-and-wellness-training

Blue Campaign Training Materials

Posters, videos, and toolkits for public education.

Website: https://www.dhs.gov/blue-campaign/materials

UNODC Education for Justice (E4J)

Academic modules on trafficking for universities and educators.

Website: https://www.unodc.org/e4j

Human Trafficking Search

Research library, news aggregation, and policy tracker.

Website: https://www.humantraffickingsearch.org

Advocacy, Volunteering, and Ethical Engagement

Freedom United

Global advocacy community campaigning against trafficking and modern slavery.

Website: https://www.freedomunited.org

Fair Trade Certified

Shop ethically and reduce demand for exploited labor.

Website: https://www.fairtradecertified.org

End It Movement

Coalition raising awareness and mobilizing public action.

Website: https://enditmovement.com

Truckers Against Trafficking

Trains transportation professionals to identify trafficking signs.

Website: https://truckersagainsttrafficking.org

Walk Free / Minderoo Foundation

Publishes the Global Slavery Index and funds anti-slavery innovation.

Website: https://www.walkfree.org

Books, Documentaries, and Further Reading

Books:

Disposable People: New Slavery in the Global Economy by Kevin Bales

Not for Sale: The Return of the Global Slave Trade and How We Can Fight It by David Batstone

Girls Like Us by Rachel Lloyd

Half the Sky: Turning Oppression into Opportunity for Women Worldwide by Nicholas D. Kristof and Sheryl WuDunn

The Slave Next Door: Human Trafficking and Slavery in America Today by Kevin Bales and Ron Soodalter

A Crime So Monstrous: Face-to-Face with Modern-Day Slavery by E. Benjamin Skinner

Documentaries:

Nefarious: Merchant of Souls (2011)

I Am Jane Doe (2017)

The Storm Makers (2014)

Trafficked (2020, National Geographic)

If You Suspect Trafficking: Immediate Steps

1. Do not confront the suspected trafficker.

2. Do not attempt to rescue the victim yourself.

3. Contact the National Human Trafficking Hotline (U.S.): 1-888-373-7888

4. If outside the U.S., contact local authorities or consult: https://www.unodc.org/unodc/en/human-trafficking

5. Document what you observe safely (location, descriptions, vehicle information) without putting yourself or the victim at risk.

Ending human trafficking is not the job of one agency, one nonprofit, or one nation. It is the shared responsibility of all

of us. Use these resources not just as a directory, but as a springboard for action.

Together, informed and equipped, we can build a world where freedom is the standard, not the exception.

Glossary of Anti-Trafficking Technology Terms

This glossary defines key technical terms used throughout the book in plain, accessible language.

Algorithm. A set of rules or instructions that a computer follows to solve a problem or complete a task. In anti-trafficking work, algorithms analyze data to detect patterns associated with exploitation.

Artificial Intelligence (AI). Technology that enables computers to perform tasks that normally require human intelligence, such as recognizing images, understanding language, and making predictions.

Augmented Reality (AR). Technology that overlays digital information, such as images, text, or alerts, onto the real-world environment, usually through a smartphone or specialized glasses.

Big Data. Extremely large datasets that are too complex for traditional analysis but can be processed by specialized tools to reveal patterns, trends, and associations.

Bitcoin. A decentralized digital currency that uses blockchain technology. Transactions are pseudonymous, meaning they are recorded publicly but not linked to real-world identities by default.

Blockchain. A decentralized digital ledger that records transactions across multiple computers. Once recorded, data cannot be altered, making it useful for transparency, accountability, and secure record-keeping.

Computer Vision. A field of AI that enables computers to interpret and analyze visual information from cameras, images, and video feeds.

Cryptocurrency. Digital currency that uses cryptography for security and operates independently of central banks. Common examples include Bitcoin, Ethereum, and Monero.

CSAM (Child Sexual Abuse Material). Any visual depiction of sexually explicit conduct involving a minor. The term is preferred over older terminology because it accurately reflects the nature of the content as documentation of abuse.

Dark Web. A portion of the internet that is intentionally hidden and requires specialized software (such as the Tor browser) to access. It provides anonymity to users and operators.

Deepfake. Synthetic media, usually video or audio, created using AI to realistically depict someone saying or doing something they never actually did.

Deep Web. The portion of the internet not indexed by standard search engines, including password-protected databases, private intranets, and academic archives. The dark web is a small subset of the deep web.

Encryption. The process of converting information into a code that prevents unauthorized access. End-to-end encryption ensures that only the sender and recipient can read a message.

Explainable AI (XAI). AI systems designed to provide human-understandable explanations of how they reach conclusions, enabling oversight, auditing, and accountability.

Facial Recognition. AI technology that identifies or verifies individuals by analyzing facial features from images or video footage.

FinCEN. The Financial Crimes Enforcement Network, a bureau of the U.S. Department of the Treasury that combats financial crimes including money laundering associated with trafficking.

Generative AI. AI systems that create new content, including text, images, audio, and video, based on patterns learned from training data. Examples include ChatGPT, DALL-E, and Midjourney.

Geospatial Intelligence. Intelligence derived from the analysis of geographic and location-based data, often using satellite imagery, GPS, and mapping technology.

Hash Matching. A technique that creates a unique digital fingerprint (hash) of an image or file, allowing systems to identify copies of known content even when modified. PhotoDNA and PDQ are widely used hash-matching technologies.

Internet of Things (IoT). The network of physical devices, vehicles, appliances, and other objects embedded with sensors and connectivity that enable them to collect and exchange data.

Machine Learning (ML). A subset of AI in which algorithms improve through experience, learning from data to make predictions or decisions without being explicitly programmed for each scenario.

Monero. A privacy-focused cryptocurrency that obscures transaction details, making it more difficult to trace than Bitcoin. It is favored by criminals seeking financial anonymity.

Natural Language Processing (NLP). A branch of AI that enables computers to understand, interpret, and generate human language, including detecting coded communication and grooming patterns.

OSINT (Open-Source Intelligence). Intelligence gathered from publicly available sources such as social media, news, business registries, and satellite imagery.

PhotoDNA. A hash-matching technology developed by Microsoft that creates unique digital fingerprints of known CSAM images, enabling platforms to detect and remove them automatically.

Predictive Analytics. The use of data, statistical algorithms, and machine learning to identify the likelihood of future outcomes based on historical data.

Privacy Coin. A cryptocurrency designed with enhanced privacy features that make transactions difficult to trace. Monero is the most widely known privacy coin.

Quantum Computing. A type of computing that uses quantum-mechanical phenomena to process information exponentially faster than classical computers.

Smart Contract. A self-executing digital contract with terms written directly into code on a blockchain. Smart contracts automatically enforce agreements when conditions are met.

Tor (The Onion Router). Free software that enables anonymous internet communication by routing traffic through a network of volunteer-operated servers, making it difficult to trace user activity.

Two-Factor Authentication (2FA). A security process requiring two forms of identification (such as a password and a phone-generated code) to access an account.

UAV (Unmanned Aerial Vehicle). Commonly known as a drone. A remotely operated or autonomous aircraft used for surveillance, reconnaissance, and delivery.

Virtual Private Network (VPN). A service that encrypts internet traffic and masks the user's IP address, providing privacy and security when browsing online.

Virtual Reality (VR). Technology that creates a fully immersive, computer-generated environment, typically experienced through a headset, that simulates physical presence in a virtual world.

About the Author

Damien Brinson is the Founder and Executive Director of Deebo Haven Center, an Orlando, Florida-based nonprofit organization dedicated to providing refuge, restoration, and resources to survivors of human trafficking. His work sits at the intersection of faith, technology, and justice, driven by a conviction that every person deserves to live free from exploitation.

Damien's path to anti-trafficking work was shaped by years of witnessing how systems meant to protect people too often fail the most vulnerable. That awareness became a calling. He founded Deebo Haven Center to fill the gaps he saw firsthand: the need for safe housing, the need for trauma-informed care, and the need for technology that serves survivors rather than surveils them.

As a writer, Damien brings a prophetic voice to nonfiction, blending investigative research with narrative urgency. *The Global Tech Revolution to End Human Trafficking* is his first book, born from the belief that the fight against modern slavery requires not just compassion but innovation, and that the people building the future need to hear from the people living in the wreckage of the present.

Damien lives in Orlando, Florida, where he continues to lead Deebo Haven Center's mission to ensure that no survivor is left without a path forward.

Connect with Damien:

Email: WeCare@deebohavencenter.org

Website: https://www.deebohavencenter.org

About Deebo Haven Center

Deebo Haven Center is a nonprofit organization based in Orlando, Florida, dedicated to combating human trafficking through direct survivor support, community education, and technology-driven advocacy. The name *Deebo* comes from a word meaning *to protect*, and that principle is at the heart of everything the organization does.

Founded by Damien Brinson, Deebo Haven Center operates on a simple but uncompromising premise: survivors of trafficking deserve more than rescue. They deserve restoration. That means safe housing, trauma-informed counseling, legal advocacy, job training, digital literacy, and long-term support systems designed to help survivors rebuild their lives on their own terms.

The Center also works to bridge the gap between the anti-trafficking community and the technology sector. Through partnerships with engineers, data scientists, and cybersecurity professionals, Deebo Haven Center advocates for survivor-centered technology design, ensuring that the tools built to fight trafficking are informed by the people who have lived through it.

Deebo Haven Center's programs include:

Survivor Support Services. Emergency and transitional housing, case management, mental health counseling, and peer mentorship for trafficking survivors.

Digital Literacy and Workforce Development. Training programs that equip survivors with technology skills, online safety knowledge, and career pathways in the digital economy.

Community Education and Prevention. Workshops, school programs, and public awareness campaigns designed to help communities recognize trafficking and respond effectively.

Technology Advocacy. Collaboration with tech companies and researchers to develop ethical, survivor-informed tools for detection, prevention, and aftercare.

To learn more, volunteer, donate, or partner with Deebo Haven Center:

Website: https://www.deebohavencenter.org

Email: WeCare@deebohavencenter.org

Location: Orlando, Florida